Primary Geography

Pupil Book 5 Change

Stephen Scoffham | Colin Bridge

Unit 1 Seas and oceans

Lesson 1: Beneath the surface

What is it like under the oceans?

We know less about the oceans than any other part of the world. People want to find out more about the animals that live in the water, how the oceans affect the weather and what happens on the ocean floor.

Exploring the oceans is difficult. There is plenty of light at the surface but below 200 metres it is almost completely dark. The weight of water is so heavy that people can only survive if they are in a submarine. It is also very cold.

Recently, scientists have discovered underwater vents. These pump fountains of boiling water and minerals into the ocean. Large numbers of animals live around the vents. Some of the animals have shells and look like crabs and shrimps. There are also huge worms that have no mouths or stomachs.

In many places the ocean floor is several kilometres below the surface.

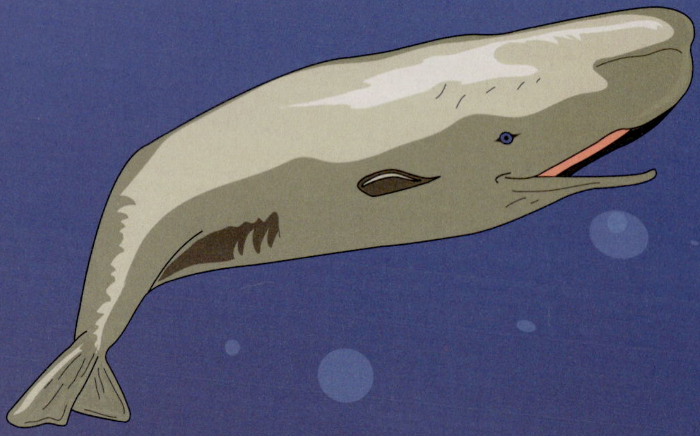

Key words

minerals
ocean floor
trench
vent

Discussion

- Why do people want to explore the oceans?
- What makes exploring the oceans difficult?
- What is the difference between deep-sea creatures and those that live near the surface?

Data Bank

- There are more volcanoes under the ocean than on dry land.
- The Marianas Trench in the Pacific Ocean is so deep (nearly 11,000 metres) that Mount Everest would fit into it.

▲ Most plants and animals live within 100 metres of the ocean surface.

▲ Below 500 metres there are unusual fish, like the Northern wolffish.

▲ Minerals from an underwater vent provide food for these deep-sea animals.

Mapwork

Make three drawings of what you might see as you go down to the ocean floor in a submarine.

Investigation

Make a class scrapbook about the oceans using the internet, newspapers and magazines.

Lesson 2: The ocean environment

What are the threats to the ocean environment?

Climate change

The oceans play an important part in creating different climates. In the Tropics the warm water evaporates bringing rain to dry land. Ocean currents also bring heat from the Equator to other parts of the world. Global warming threatens to change these patterns.

▼ Storm clouds gather over the ocean.

Key words

climate
Equator
global warming
ocean currents
Tropics

Atlanti
Ocear
Panama
Canal
Equator
Pacific
Ocean
N

Discussion

- In what ways are oceans threatened?
- Why do threats to the ocean environment matter?
- How can the oceans be protected?

Mapwork

Draw a map of the Arctic Ocean. Add notes about the way it is threatened or changing.

Over fishing

Many people like to eat fish but supplies are getting harder to find. This is because modern fishing boats drag long nets through the water catching everything in their path. In the Atlantic Ocean, stocks of herring and cod are very low. Around Antarctica so many whales have been killed that they are in danger of extinction.

▶ Sorting through the catch.

Shipping

Most of the goods that are moved around the world are carried by ships. Tankers and bulk carriers are loaded with oil, coal, iron ore and other heavy cargoes. Some of the most important shipping routes are around the coast of Europe, Southeast Asia and the USA. If ships collide or run aground it can cause terrible pollution.

▶ A container ship with thousands of containers.

rctic Ocean
Suez Canal
Pacific Ocean
Indian Ocean
Southern Ocean

Pollution

The oceans are vital to the health of the planet. However, scientists are concerned that they are being used as rubbish dumps. Plastic and garbage are building up where currents are weak. Sea water is getting more acidic. In the Arctic Ocean old nuclear submarines have been left to rot on the sea floor.

▼ A dolphin leaping through the waves.

Investigation

Design a poster about threats to the ocean environment.

Lesson 3: Learning about seas

What is a sea?

Around the edge of the oceans, there are places where the water is quite shallow. These are known as seas. The beaches and shore are often popular with visitors who come to enjoy the seaside.

Like oceans, seas are important habitats for fish and other animals. In some parts of the world, oil, gas and minerals have been found in rocks under the seabed. All these resources are very valuable.

▲ A coral reef in the Red Sea.

▲ Platforms are used to pump oil from under the seabed.

Discussion

- How is a sea different from an ocean?
- Why are seas so useful?
- What impact have people had on the North Sea?

Mapwork

Using an atlas, make a list of seas around the world.

Key words

fish stocks
oil platform
resort
resources
shore
wind farm

▲ Scarborough beach and seaside resort.

The North Sea

The North Sea is one of the busiest seas in the world. Thousands of ships cross it every day. Fifty million people live around the shores or the rivers which flow into it.

The North Sea is home to many seals, sea birds, fish and sea creatures. However the water has become polluted. Countries around the North Sea have now passed laws to try and make the water cleaner. This is starting to improve the environment.

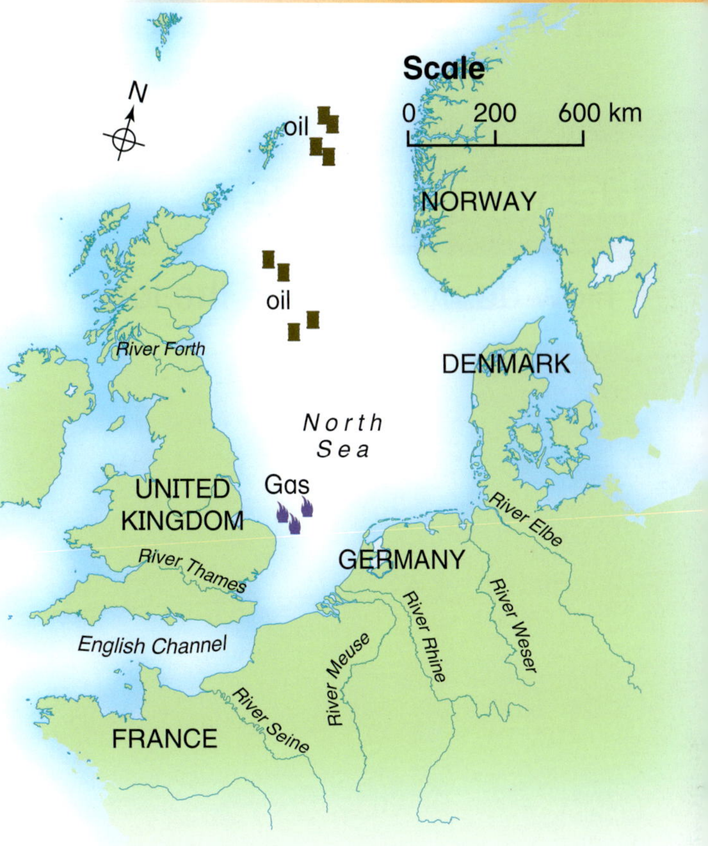

Scale
0 200 600 km
N
oil
oil
NORWAY
DENMARK
River Forth
North Sea
UNITED KINGDOM
Gas
River Elbe
GERMANY
River Thames
River Weser
River Rhine
River Meuse
English Channel
River Seine
FRANCE

- The North Sea contains three per cent of world fish stocks.
- There are several hundred oil and gas platforms in the North Sea.
- On average the North Sea is around 100 metres deep.
- Increasing numbers of wind farms are being built in the North Sea to produce clean energy.

UNITED KINGDOM
DENMARK
0 m

▲ Most parts of the North Sea are very shallow.

Investigation

Find out about the different plants and creatures in the North Sea for a class display.

Summary

In this unit you have learnt about:

- what lies beneath the surface of the ocean
- how oceans are threatened
- conserving seas and oceans.

Unit 2 Wearing away the land

Lesson 1: Rivers in action

How do rivers shape the land?

Key words

channel
deposition
erosion
reservoir
river bank
transportation
water cycle

▲ This canoeist is being carried along by the force of the water. All he has to do is steer clear of the rocks!

Data Bank

- The way that water moves round the world is called the water cycle.
- The amount of water in the world always stays the same – no new water will ever be formed.

▲ **Erosion:** Rivers cut into the land creating valleys with steep sides.

As streams and rivers flow downhill, they remove tiny pieces of rock on the river bed. They also eat into the earth banks on either side of the channel. The tiny particles of rock and earth bounce and scrape along the river bed wearing it away even more. This shapes and moulds the land over thousands and thousands of years.

A lot of the material which is carried along by the water is dropped somewhere else. Some of it slowly builds up into banks of sand, mud and gravel in the middle of the river. In other places, the material is dropped in lakes and reservoirs. Over long periods of time, these fill up and turn into dry land.

▲ **Transportation:** Flood waters are so powerful they can carry rocks, boulders and whole trees downstream.

▲ **Deposition:** Rivers drop gravel and mud which build up in banks.

Discussion

- What is the water cycle?
- How do rivers wear away the land?
- How do rivers build up the land?

Investigation

- Make a drawing and write a sentence about three special words on this page in your geography notebook.

Lesson 2: Preventing flood damage

How can we control rivers?

The Mississippi River is over 3000 kilometres long. It flows southwards across the United States of America, draining half the country. The Mississippi is an important route for shipping.

In the past, the Mississippi was very shallow and used to flood after heavy rain. The river was also constantly changing its channel as it meandered towards the sea.

A hundred years ago, a team of river engineers made the channel deeper to help shipping. They also built banks to protect nearby farms, towns and factories from floods.

▲ The Mississippi was a dangerous river for shipping because it was very shallow and there were a lot of sandbanks.

CANADA
ROCKY
MOUNTAINS
Great Lakes
Missouri
Chicago
New York
Atlantic
Ocean
UNITED STATES
OF AMERICA
St Louis
Mississippi
Pacific
Ocean
New Orleans
MEXICO
Key
Over 1000 metres
200-1000 metres
0-200 metres
Scale
0
600
1200 km

Dykes

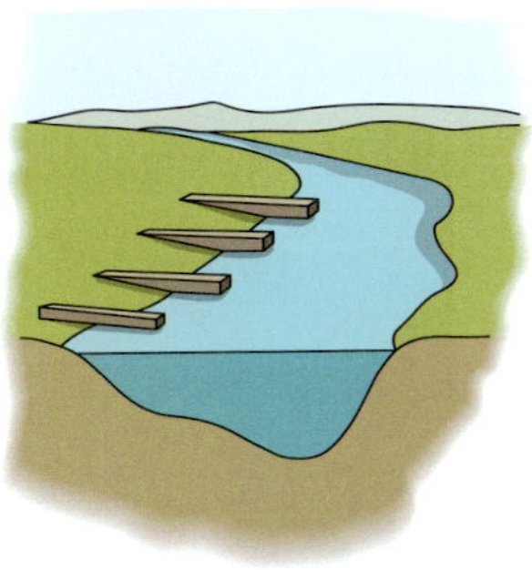

Dykes along one side of the river force the water to cut a deeper channel on the opposite side.

Levees

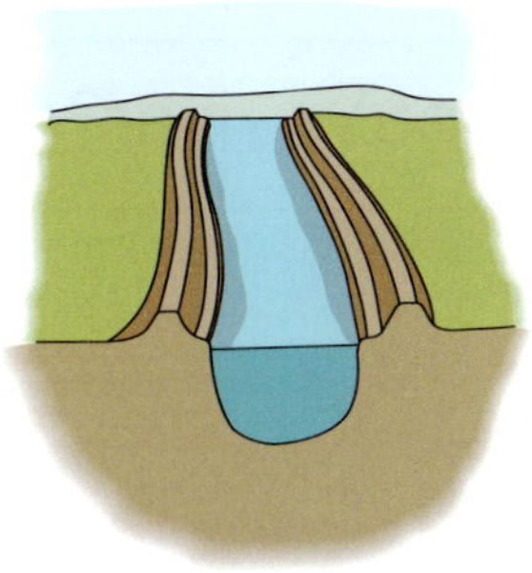

Huge earth and clay banks called levees hold back the flood water.

Cut-offs

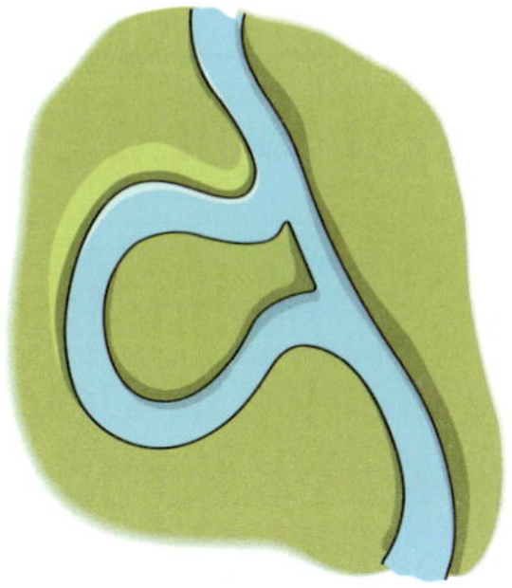

New channels cut off some of the meanders so the water can flow faster.

Boxes

The sides and bottom of the channel are lined with concrete boxes to make them stronger.

Today there is a system of dykes, levees and dams along the Mississippi. The channel has been straightened and lined with concrete boxes. However, serious floods still happen. Some people question whether it will ever be possible to tame the Mississippi. They think the river should be allowed to flood as it did in the past and that the levees and boxes only trap the water and make things worse.

Mapwork

Draw your own map of the Mississippi and its tributaries.

Investigation

Write a report about a flood in your own area. This might range from a burst pipe at home to a disaster that affected everyone. Include drawings and pictures if you can.

Key words

channel
dyke
flood
levee
meander
tributary

▼ A flooded city in Missouri.

Lesson 3: Finding out about rivers

What data is needed to find out about a river?

Children from St Mark's Primary School went to a Field Study Centre in Devon (Southwest England). As part of their work on the environment they found out about a nearby river. The children tried to answers these questions.

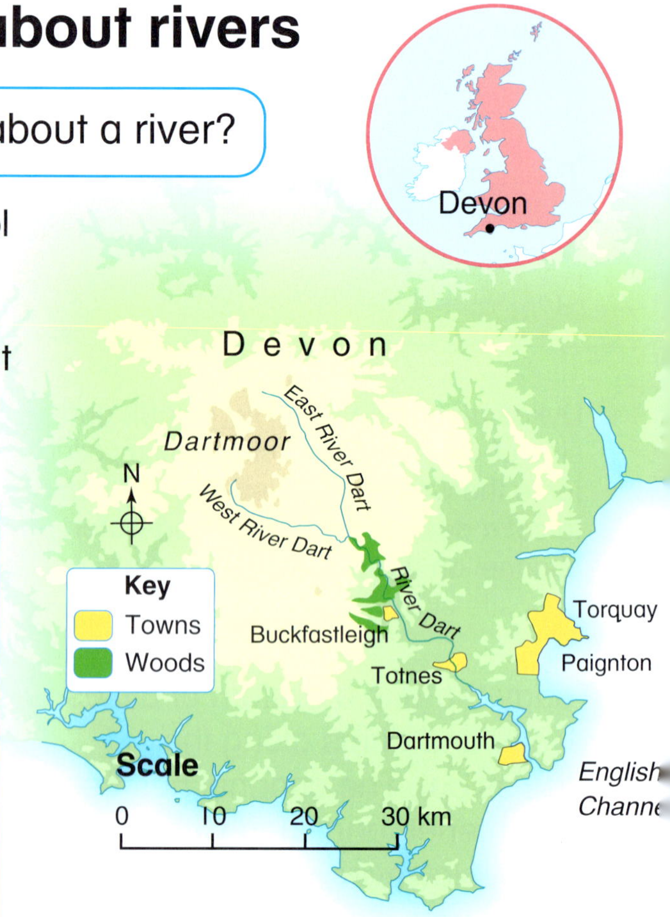

Devon
Devon
Dartmoor
East River Dart
West River Dart
River Dart
N
Key
Towns
Woods
Buckfastleigh
Totnes
Torquay
Paignton
Dartmouth
English Channe
Scale
0
10
20
30 km

River survey

1. In which compass direction does the river flow?
2. How wide is the river at three different places?
3. Is the river bed flat between the banks?
4. Are there any particles being carried along by the water?
5. Does the water flow at the same speed in middle of the river as it does as the edges?
6. What plants grow on the river bank?
7. Is there any evidence of fish or animal life?
8. Are there any clues that the water level changes?

Discussion

- Why did the children go to the Field Study Centre?
- Which question in the list do you think is the most important?

▼ The children used this equipment when they studied the River Dart.

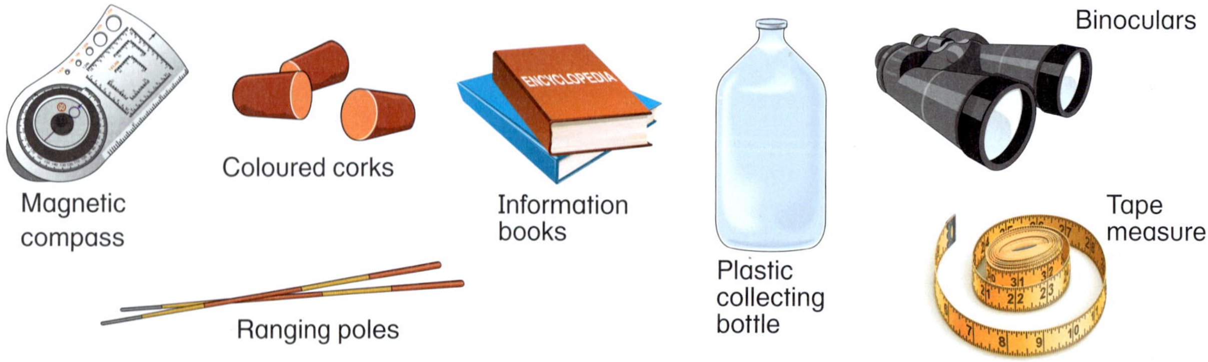

ENCYCLOPEDIA
Magnetic compass
Coloured corks
Information books
Plastic collecting bottle
Binoculars
Tape measure
Ranging poles

▲ Taking river measurements with a tape.

Our River Study

We spent Tuesday at the river. We had to measure carefully and make observations using a compass. We found that the river flowed from north to south. In places it was three metres wide. The ranging pole showed the water was half a metre deep. It was flowing quite fast. We could see water weed and animal burrows along the bank.

When the children returned to the Study Centre, they recorded their results on a computer data file. They also looked at local maps to trace the route of the river from source to mouth. Then the children also found out more about the plants and animals which they had seen on the river bank. If there is a river near your school, you could make a similar study.

Key words

environment
magnetic compass
particle
ranging pole

Mapwork

Using a local map, locate and name streams and rivers in your area.

Investigation

Decide which piece of equipment the children would need to answer each survey question.

Summary

In this unit you have learnt:

- that rivers are a major influence on the landscape
- how people try to control rivers
- how to study a river.

Unit 3 The seasons

Key words

cycle
pattern
season
temperature

Lesson 1: Changing seasons

What are the seasons?

Over the year there is a pattern to the weather depending on the season. In winter, the weather is often cold and the days are dark and short. In summer, the weather is much warmer and the days are long and bright. Spring is the time when plants begin to grow and birds build their nests. Fruit and other crops are harvested in the autumn.

The changing seasons give a pattern to our lives. They affect the clothes we wear, the things we do and the places we visit.

Discussion

- Which season is shown in each of the photographs on page 15?
- Which is the coldest and warmest month according to the temperature chart?
- How do the seasons affect people and plants?

Investigation

- Cut out a circle of card to make a seasons dial. Add drawings and notes for each of the four seasons.

▼ Changes in sunshine and temperature affect the life cycle of animals and plants.

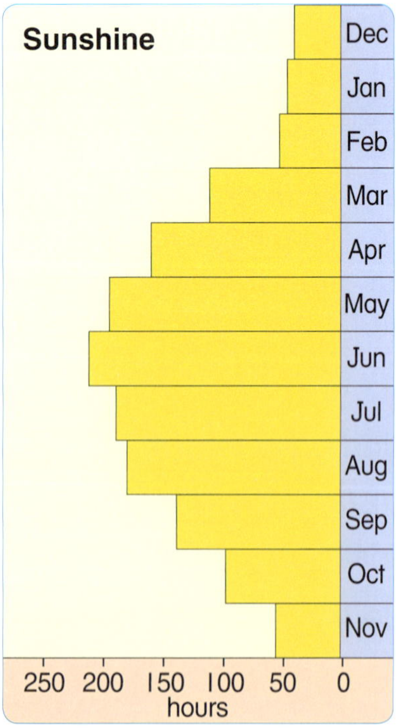
Sunshine
Dec
Jan
Feb
Mar
Apr
May
Jun
Jul
Aug
Sep
Oct
Nov
250 200 150 100 50 0
hours

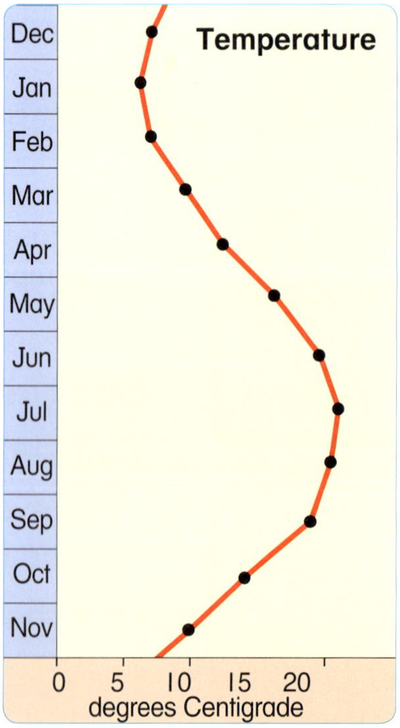
Temperature
Dec
Jan
Feb
Mar
Apr
May
Jun
Jul
Aug
Sep
Oct
Nov
0 5 10 15 20
degrees Centigrade

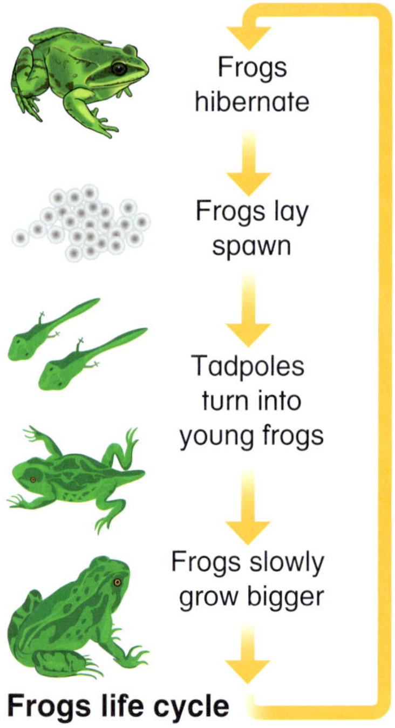
Frogs hibernate
Frogs lay spawn
Tadpoles turn into young frogs
Frogs slowly grow bigger
Frogs life cycle

A

B

C

D

Lesson 2: Seasons worldwide

Do all places have the same seasons?

Key words

climate
Mediterranean
monsoon
season
tropical

In the United Kingdom we have four seasons which each last three months. Some other parts of the world have a different pattern of seasons. This affects how people live and the crops they can grow.

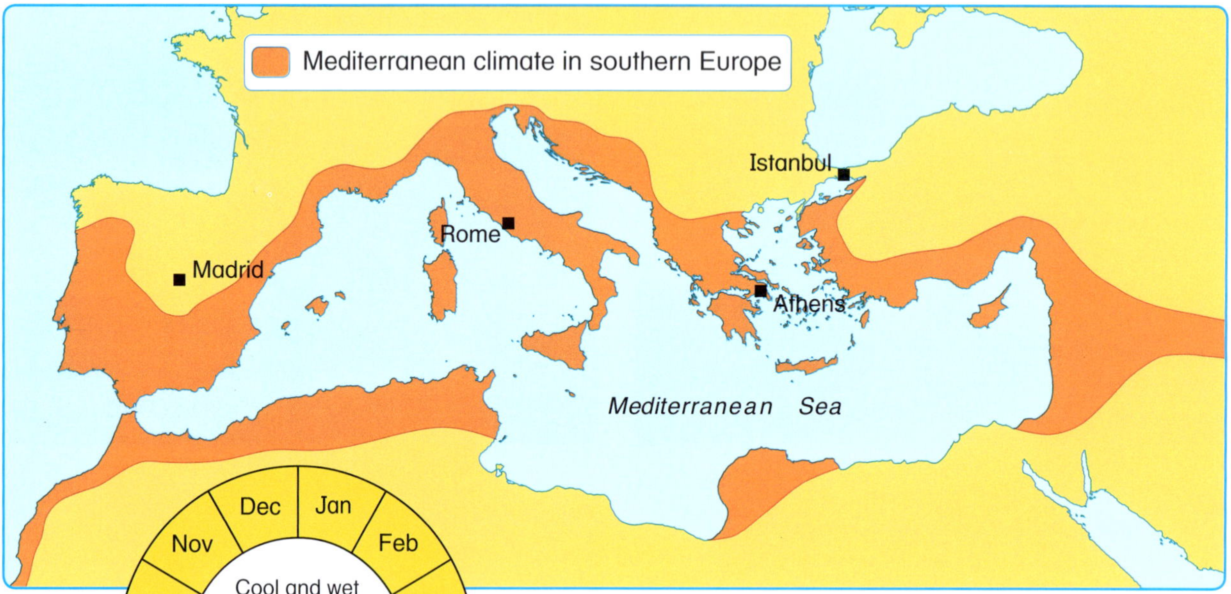
Mediterranean climate in southern Europe
Istanbul
Rome
Madrid
Athens
Mediterranean Sea

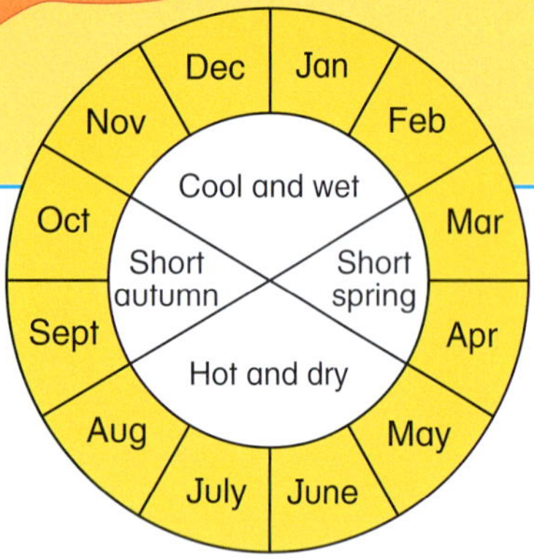
Dec
Jan
Feb
Mar
Apr
May
June
July
Aug
Sept
Oct
Nov
Cool and wet
Short spring
Hot and dry
Short autumn

▲ A Mediterranean landscape in Italy.

Southern Europe has a Mediterranean climate

Summer in the Mediterranean.

"Summer is a time of scorching heat. The countryside is alive with the humming of insects and the crackling of dry grass. In the fields the crops are ready to harvest.

During the day the whole land is flooded with light. The glare of the sun is thrown back from the white rocks. The streams dry up and only the tough plants can survive in the heat."

(Adapted from *Mehmet My Hawk* by Yashar Kemal, HarperCollins Publishers)

Discussion

- How long do seasons last in the UK?
- How does the Mediterranean climate differ from the UK?
- What is the effect of the monsoon?

Mapwork

Using an atlas make a list of other parts of the world that have Mediterranean and monsoon (tropical) climates.

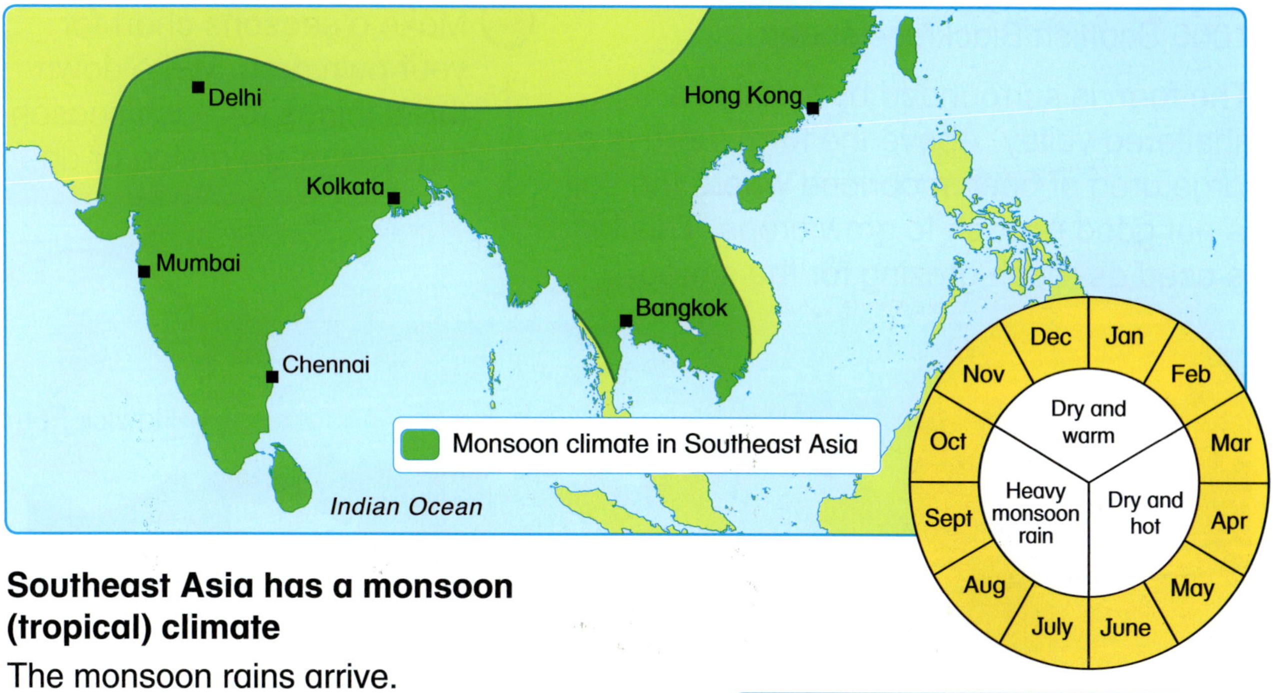

Delhi
Hong Kong
Kolkata
Mumbai
Bangkok
Chennai
Monsoon climate in Southeast Asia
Indian Ocean
Dec
Jan
Feb
Mar
Apr
May
June
July
Aug
Sept
Oct
Nov
Dry and warm
Dry and hot
Heavy monsoon rain

Southeast Asia has a monsoon (tropical) climate

The monsoon rains arrive.

"A hot wind blew through our bungalow day and night from the huge open plain. Then the clouds began to bank up and bank up and there was an unbearable feeling of pressure.

The rains came down with terrific force, such as you hardly ever see in Europe. This would probably go on for two or three days and the whole area round the houses turned green. An extraordinary life burst out."

(Adapted from *Plain Tales of the Raj* edited by Charles Allen, Futura Publications)

▲ Monsoon rains are vital for crops such as rice. One thousand million people depend on rice grown during the monsoon for their food.

Investigation

Write a description of the weather to match one of the seasons in the UK. Read it to someone else and ask them to guess which season you have described.

Lesson 3: Seasonal influences

How are farmers affected by the seasons?

Key words

harbour
heat wave
moorland
Pennines
resort

Hawick Farm is in the Pennines in northern England. The farmer has 50 cattle and 1000 Scottish Blackface sheep.

The farm is surrounded by fields in a sheltered valley. Above the farm there is a large area of open moorland where the soil is not good enough to grow crops. This land is used as rough grazing for the sheep.

Investigation

Make a seasons chart for your own area. Write down three things for each season that you might notice or do.

Hawick Farm

▲ Cows grazing in the fields around the farm.

Mapwork

Plan a short walk where people might see spring flowers e.g. in parks, woods and orchards. Draw a sketch map of the route and add notes about the main points of interest.

	Winter	Spring	Summer	Autumn
Sheep	Sheep fed by farmer during snowy weather.	Sheep brought down from the moors for lambing. Lambs marked and tails removed.	Sheep dipped to stop disease. Wool clipped and sold. Sheep taken back to the moors.	Sheep dipped again. Mating time.
Cattle	Cattle fed on hay and turnips in the barns.	Calves born. Some cattle sold at market.	Mating of cows.	Cattle put in the fields after haymaking.
Other jobs	Repairs to stone walls.	Muck spread in the fields to help the grass grow.	Haymaking.	Drains repaired in the fields.

Data Bank

- The amount of electricity used in houses nearly doubles in cold winter weather.
- In a summer heat-wave sales of ice cream can increase fourfold.
- In the UK around two million tonnes of salt are used each year to free roads from snow and ice.

How are seaside resorts affected by the seasons?

Whitby is a town on the coast of Yorkshire in England. It has a busy fishing harbour and a long sandy beach. Many people go to Whitby for their holidays. Some people visit the old abbey.

▼ The number of visitors to Whitby Abbey.

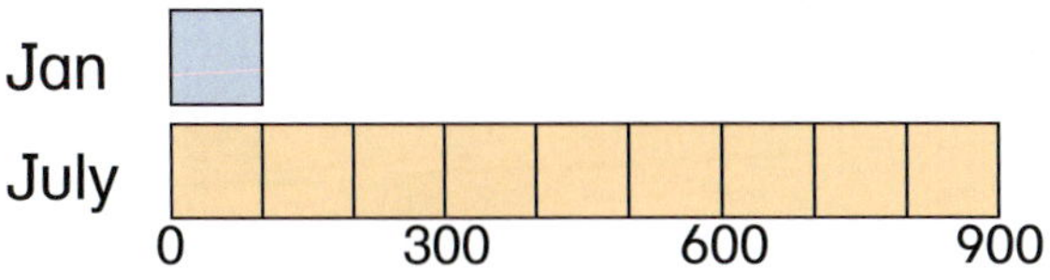
Jan
July
0
300
600
900

▲ Tourists catching fish and crabs in Whitby harbour.

Summer

In summer Whitby is crowded with visitors who come from all parts of the UK and abroad.

- Beach busy with tourists.
- Boat trips.
- Steam railway.
- Museum open every day.

Winter

In winter Whitby is much quieter, particularly when the wind brings rain and snow from the North Sea.

- Beach empty.
- Sea too rough for boat trips.
- Railway line repaired.
- Museum only open at weekends.

Discussion

- ☐ What are the Pennines?
- ☐ How does Whitby change between summer and winter?
- ☐ Which is your favourite season?

Summary

In this unit you have learnt:

- how the seasons are different
- about seasons around the world
- how people are affected by the seasons.

Unit 4 Cities

Lesson 1: Describing cities

What are cities like?

▼ Skyscrapers in New York City, USA.

Key words

city centre
settlement
skyscraper
suburbs
vandalism

Discussion

- Why do people want to visit city centres?
- Why are there suburbs around cities?
- Why do you think cities are getting larger?

Cities are the largest of all settlements. They are busy, crowded places with hundreds of thousands of homes.

During the day, people come to the city centre to buy things in shops and work in office blocks. As night falls, people leave work and go to restaurants, theatres, cinemas and clubs. The streets are full of bright lights from the shops and advertisements. Away from the centre, it is quieter. The suburbs spread out into the countryside. There is more space there for houses and gardens, shops and parks.

Cities are connected to other places by high-speed trains, motorways and airports. The roads and underground trains are often crowded. Noise and fumes can be a problem. Some areas suffer from vandalism.

Data Bank

- The Tacoma Building in Chicago, USA (1889) was one of the world's first skyscrapers.
- Hong Kong has more skyscrapers than any other city in the world.

There are lots of jobs in the city centre.
All the fumes, from the traffic are bad for my health.
We like living in the city because there are lots of things to do in the evening.
It's very noisy and crowded where we live.
I have come to the city to study at university.
There are no open spaces to play ball games where I live.

Investigation

Draw a picture of yourself. Add a speech bubble saying what you think about city life.

Mapwork

Working from an atlas make a list of ten cities in the UK.

Lesson 2: World cities

Key words

city
countryside
population

How are cities changing?

Around the world, many cities are getting bigger as people move in from the countryside. Some people are attracted to cities because they think they will find a better life there. Others are forced to move out of their villages by war, drought and famine.

City	Millions of people	
	1980	2020 (estimate)
Tokyo, Japan	29	39
Delhi, Inda	6	29
Shanghai, China	6	26
Mumbai, India	9	24
Mexico City, Mexico	13	23
New York, USA	16	22
São Paulo, Brazil	12	22
Beijing, China	5	21
Dhaka, Bangladesh	3	20
Karachi, Pakistan	5	18
Kolkata, India	9	17
Lagos, Nigeria	3	16

Discussion

- Which will be the three largest cities in the world in 2020?
- Why do cities attract people?
- What is special about the city nearest to you?

London
Paris
EUROPE
NORTH AMERICA
New York
ASIA
Beijing
Tokyo
Atlantic Ocean
Cairo
Delhi
Dhaka
Shanghai
Karachi
Mexico City
Mumbai
Kolkata
AFRICA
Pacific Ocean
Lagos
SOUTH AMERICA
OCEANIA
Indian Ocean
São Paulo
Buenos Aires
Sydney
ANTARCTICA

▲ Cities around the world.

Investigation

Using the data from the world cities table above draw a bar chart of the world's city populations.

The story of New York

New York is on the east coast of the United States of America. It began as a trading post set up by Dutch sailors nearly 400 years ago. Now it has a population of 20 million people speaking 800 languages and is an important centre for world trade and finance.

New York is famous for the Statue of Liberty and the Empire State Building. Central Park is a popular meeting place. The museums of modern art and natural history attract many visitors. In the evening there are shows to see in the 40 theatres along Broadway.

New York enjoys sunshine on most days in the year but January can be very cold and sudden deep snow is not unusual.

Mapwork

Using the internet, make your own simple sketch map of central New York.

Data Bank

- The Empire State Building has 102 floors and 6500 windows.
- In New York, people spend over one billion dollars on theatre tickets each year.
- There are 70 000 helicopter flights over New York each year.

Central Park
Empire State Building
Statue of Liberty
Broadway

▲ Central New York.

Lesson 3: The story of London

Key words

crossing point
population
route map
Thames
underground

How has London grown and changed?

London was built at a crossing point on the River Thames. Low hills on the north bank of the Thames provided a dry site above the marshes. Ships could sail up the river carrying goods from other parts of Europe.

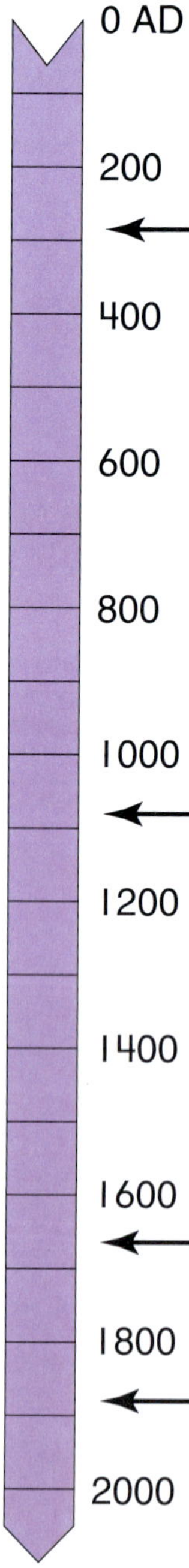
0 AD
200
400
600
800
1000
1200
1400
1600
1800
2000

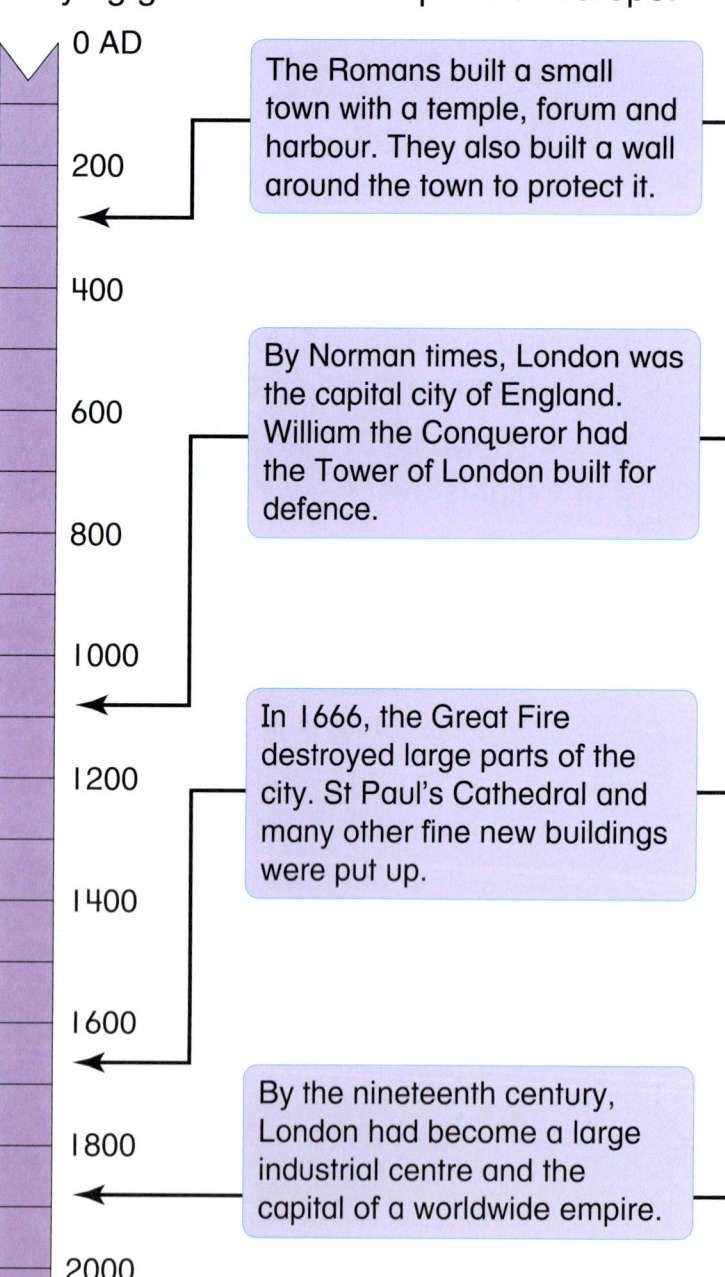

The Romans built a small town with a temple, forum and harbour. They also built a wall around the town to protect it.

Ancient Roman wall ruin.

By Norman times, London was the capital city of England. William the Conqueror had the Tower of London built for defence.

The Tower of London.

In 1666, the Great Fire destroyed large parts of the city. St Paul's Cathedral and many other fine new buildings were put up.

St Paul's Cathedral.

By the nineteenth century, London had become a large industrial centre and the capital of a worldwide empire.

The Houses of Parliament.

British Museum
Holborn
Sir John Soane Museum
REGENT STREET
OXFORD STREET
Soho
Eros
London Transport Museum
STRAND
Mayfair
National Gallery
Trafalgar Square
Strand
Charing Cross
PICCADILLY
St. James's
WHITEHALL
London Eye
The Green Park
St. James's Park
Downing Street
The Cenotaph
Waterloo
London Aquarium
Buckingham Palace
Westminster
STREET
Big Ben & Houses of Parliament
Westminster Abbey
Victoria
VICTORIA
Westminster Cathedral
Archbishops Park
Victoria
Lambeth
VAUXHALL BRIDGE ROAD
Tate Britain

▲ A tourist map for children showing landmarks in central London.

Data Bank

- London was the first city in the world to have underground trains.
- There are 24 bridges spanning the Thames in London.
- Around a quarter of the people who live in London were born abroad.

Mapwork

Draw a map of a walk through London linking four different landmarks.

Investigation

Find out one thing about each of the cities marked on the UK map.

Glasgow
Edinburgh
Newcastle
Belfast
Leeds
Manchester
Birmingham
Cardiff
Bristol
London

Discussion

- Why was London built?
- Why did London grow so much in the nineteenth century?
- What would you like to see and do if you visited London?

Summary

In this unit you have learnt about:

- how cities are different from other places
- how cities are changing
- how to describe a city.

Unit 5 Jobs

Lesson 1: Making things

Where are things made?

Most of the things that we eat, wear and use each day are made in factories. Factories can be very small places with one or two workers or very large places with thousands of workers.

Factories use machines to make things quickly and cheaply. Machines can make large numbers of things in exactly the same way each time.

All factories need:

- Workshops, offices and storage space.
- A main entrance, a delivery area and car park.
- A power supply for machines.
- A way of getting rid of waste.

Key words

dough	storage
ingredients	waste
Input-output	workshop
raw materials	

Discussion

- What can factories do?
- What do factories need?
- What are the differences between a factory and your school? Is anything the same?
- What can you see around you that might have been made in a factory?

Offices
Chimney
Workshop
Boiler
Storage space
Car park
Delivery area
Main entrance

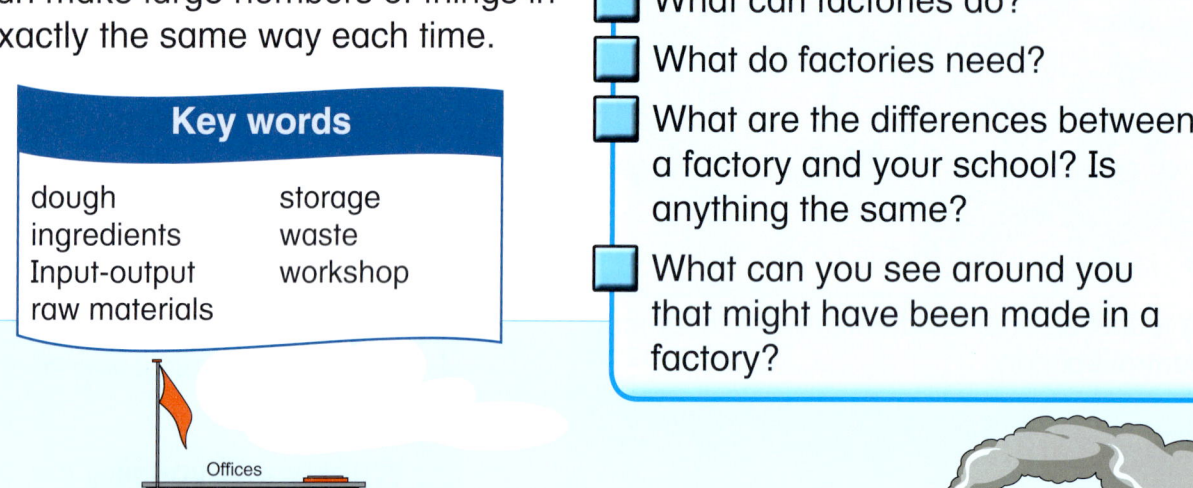

▲ Modern factory.

How do factories work?

Before it can make anything a factory needs raw materials. These materials are either dug out of the earth or produced by farmers and fishermen. Factories turn these raw materials into the goods we see in shops.

Investigation

Make Input-output diagrams for making a pencil and a carton of apple juice.

Making bread
Input-output diagram

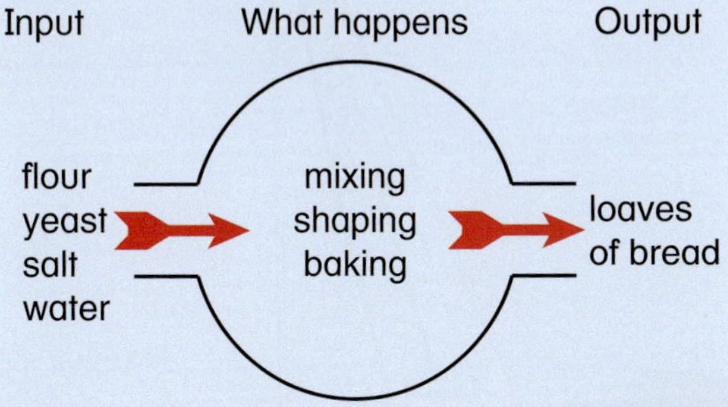
Input
What happens
Output
flour
yeast
salt
water
mixing
shaping
baking
loaves
of bread

▲ Inside a large bakery.

Bread is made from flour, yeast, salt and water.

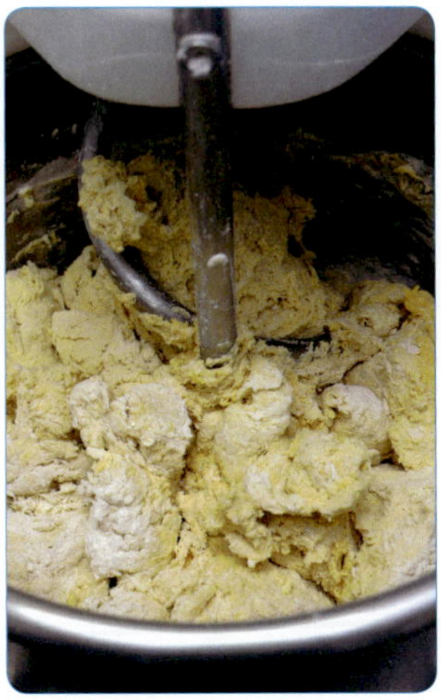

The ingredients are mixed to make dough.

The dough is shaped into loaves and baked in the oven.

The bread is sliced, and put into crates.

INPUT → **WHAT HAPPENS** → **WHAT HAPPENS** → **OUTPUT**

Lesson 2: Different jobs

Key words

bulk carrier
harbour
lifeboat
skill

How do people earn a living?

Most people who do a job are paid for what they do. The amount they are paid depends on their skills, their qualifications and the number of hours they work.

There are different types of jobs. Some people work in offices, collecting information, adding up figures or making decisions. Others work outdoors, operating machinery or running farms.

Joan Lovell works in a factory. She cleans and boils crabs, lobsters and other shellfish so they are ready to be packed to send to the shops.

Bill Shaw is a fisherman.

Discussion

- What is a harbour?
- Which job is the most important?
- Which job do you think is most interesting?

Investigation

What single special skill does each job require? Show your answers in a table.

Mapwork

Make a plan of the harbour from the clues in the picture.

Jack Seymour is a crane driver. He unloads cargoes from the boats.

Andrew Knight is the harbour master. He gives permission for boats to come in and out of the harbour.

Steven Bell is the coxswain (captain) of the lifeboat.

Winston Hayes is an engineer. He checks and maintains all the machinery in the harbour.

Susan White drives a bulk carrier. It carries gravel chippings from the harbour to local road works.

Maria Archer keeps records and accounts in the harbour master's office.

Lesson 3: Types of work

Key words

primary
retirement
secondary
tertiary

What are the different types of work?

There are three main types of work.

In the past most people in the UK worked in primary activities. Now tertiary activities have become much more important.

- **Primary activities** – collecting and using materials from the Earth's surface.
- **Secondary activities** – making things from natural resources.
- **Tertiary activities** – providing a service for people to use.

Primary

Collecting raw materials

e.g. farming, fishing, forestry, mining.

Secondary

Making things

e.g. factory work, making goods, making roads, putting up buildings.

Tertiary

Providing a service

e.g. shop and office work, driving vehicles, caring for people, entertaining people.

Data Bank

Pie chart of UK work force

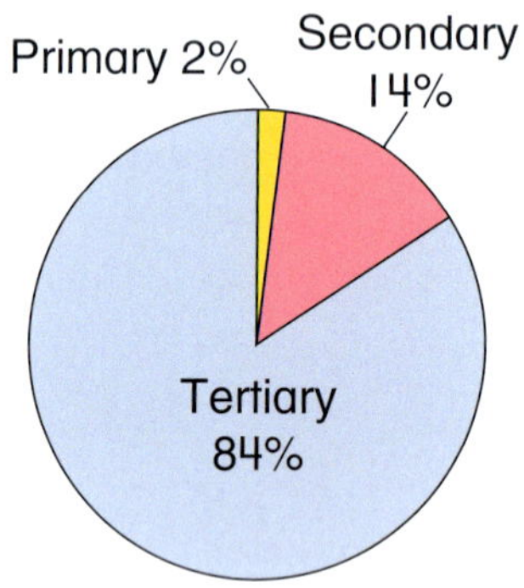
Primary 2%
Secondary 14%
Tertiary 84%

- In the UK women are still paid less than men for the same work.
- Many people are out of work and cannot find jobs.
- The retirement age is increasing from 65 to 68 or 70 years.

Discussion

- What categories do the jobs which adults do in your school fall into?
- Which type of work do you think is hardest?
- Why are fewer people needed in primary activities?

In one school children used a local newspaper to make a survey of jobs in their area. They copied some of the advertisements and put them into a scrapbook using separate pages for primary, secondary and tertiary work.

Investigation

Make a scrapbook of your own using the advertisements from this page together with ones from a local paper.

VAN DRIVER
required by international food company.
We are looking for a committed and reliable person for this permanent job.

Chemist
Local company seeks bright young scientist for laboratory work. Duties include quality control and testing.

St Stephen's Junior School
Secretary
Required for this busy, popular friendly school. 35 hours per week. Office experience and knowledge of computers an advantage.

WE NEED YOU
STRAWBERRY PICKERS
(from mid May)
Good rates of pay.
Transport available.

City centre pub requires young chef

Printer
Full time 9.00am to 5.30pm. Must be able to work on their own. Knowledge of Apple Mac computer an advantage.

ASSEMBLY LINE WORKERS
Modern company has four full time positions. Light work, training provided.

Leisure Homes
require a Swimming Pool Attendant. Qualification required.

Sales Manager
needed for expanding local business. Must be willing to travel.

Coach Driver needed for summer season. Licence required.

Office Person
Busy office. Good communication skills essential.

Nursing Home requires a Deputy Matron
This is a new post which would be ideal for an experienced nurse.

Mapwork

Make a plan of your house and garden or make up one of your own. Now add annotations to show where primary, secondary and tertiary activities happen.

Summary

In this unit you have learnt:

- what factories do
- how jobs are linked together
- about how jobs can be grouped.

Unit 6 Pollution

Lesson 1: Damaging the environment

What causes pollution?

When fumes, noise and waste cause damage to the environment it is called pollution. Some natural events cause pollution. For example, if a volcano erupts it pollutes the air with large quantities of dust and poisonous gas. However, many pollution problems are caused by people.

Discussion

- What causes pollution?
- Is all pollution caused by people?
- Think about different types of pollution. Which do you think are the most serious?

Key words

chemicals
nuclear waste
pollution
volcano

▼ It took 87 days to stop the oil spill at the Deepwater Horizon oil rig after an explosion in 2010.

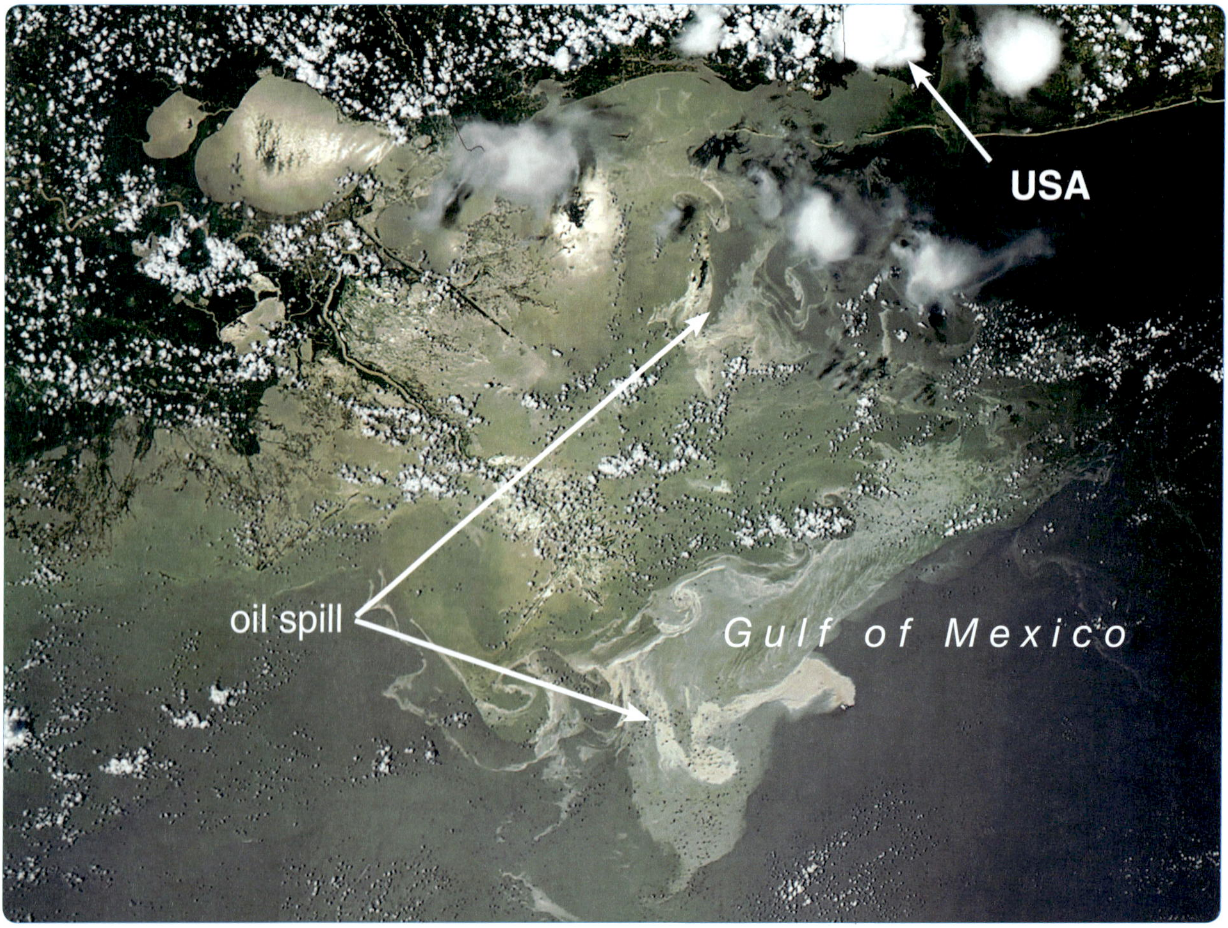
USA
oil spill
Gulf of Mexico

How do we cause pollution?

In the summer holidays we often go out for the day in the car.

We all like to travel but cars clog up the roads and put poisonous fumes into the air.

I like getting presents at Christmas.

We all enjoy getting presents but some of the factories which make toys put chemicals into rivers.

I like eating crisps.

We all need to eat but the rubbish from food packets has to go somewhere.

Air pollution

Water pollution

Land pollution

How long does pollution last?

Pollution does not last forever. Fruit and vegetable peelings will rot in a few weeks. Paper and cardboard disappear in about a year. Metal lasts for many years because it rusts slowly.

▶ This is the order in which things decay.

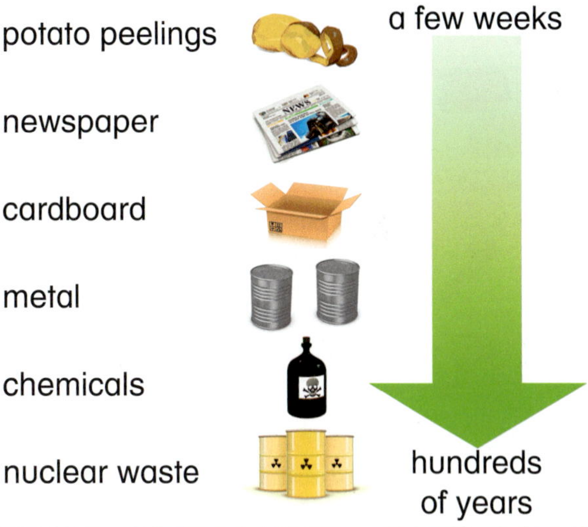
potato peelings
newspaper
cardboard
metal
chemicals
nuclear waste
a few weeks
hundreds of years

Mapwork

Make a map to show clean and dirty areas in your school building or local surroundings.

Investigation

Choose six objects in your home or classroom. Decide how long each one would take to decay and show your results in a chart.

Data Bank

- On average, each person in the UK throws away their own body weight in rubbish every seven weeks.
- A plastic bag can take up to 100 years to decompose.
- Nuclear waste can take between 10,000 and a million years to decay.

Lesson 2: 'Green living'

Key words

dam
recycle
renewable resources
solar panels
turbine
wind farm

How can we reduce pollution?

We can all help to reduce pollution. We could buy fewer things which would save resources. We could also make sure that the things we no longer need are recycled.

Recently the government has passed new laws to protect the environment. The trouble is that it takes a long time for people to change their habits.

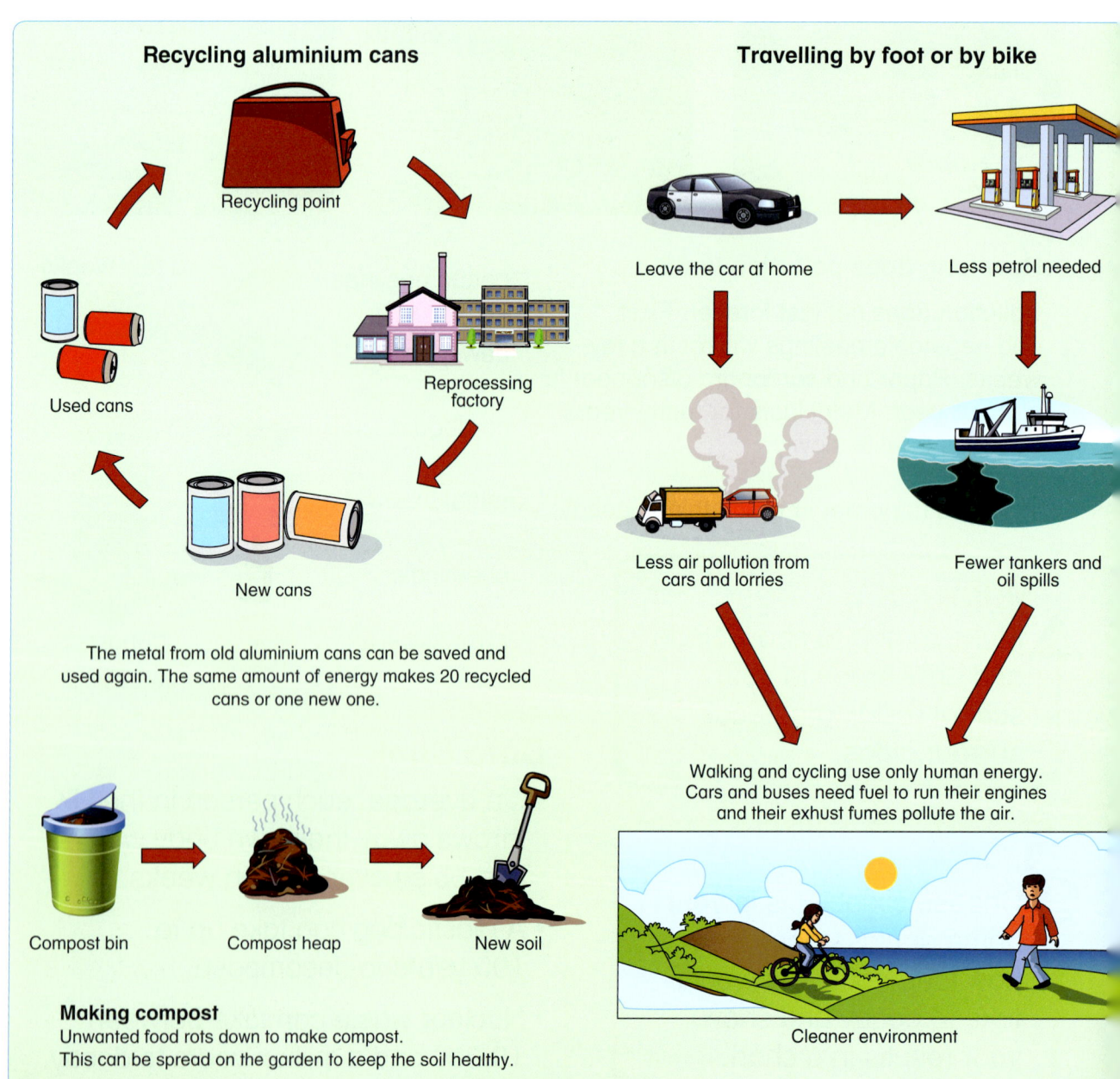

Recycling aluminium cans

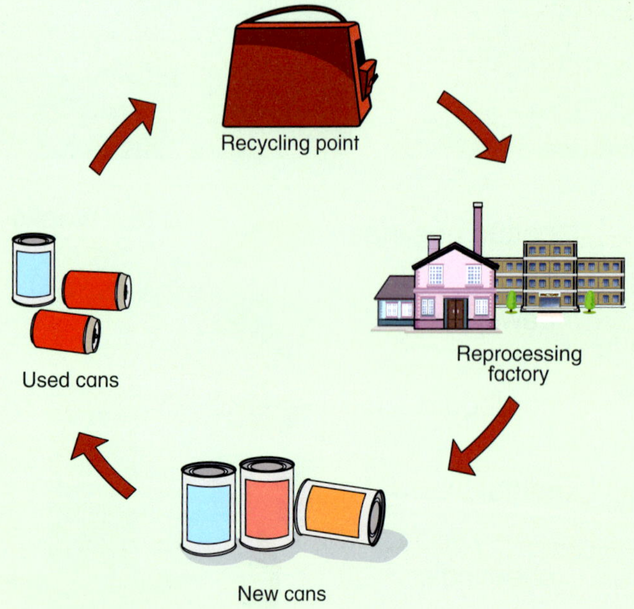
Recycling point
Reprocessing factory
New cans
Used cans

The metal from old aluminium cans can be saved and used again. The same amount of energy makes 20 recycled cans or one new one.

Compost bin
Compost heap
New soil

Making compost
Unwanted food rots down to make compost.
This can be spread on the garden to keep the soil healthy.

Travelling by foot or by bike

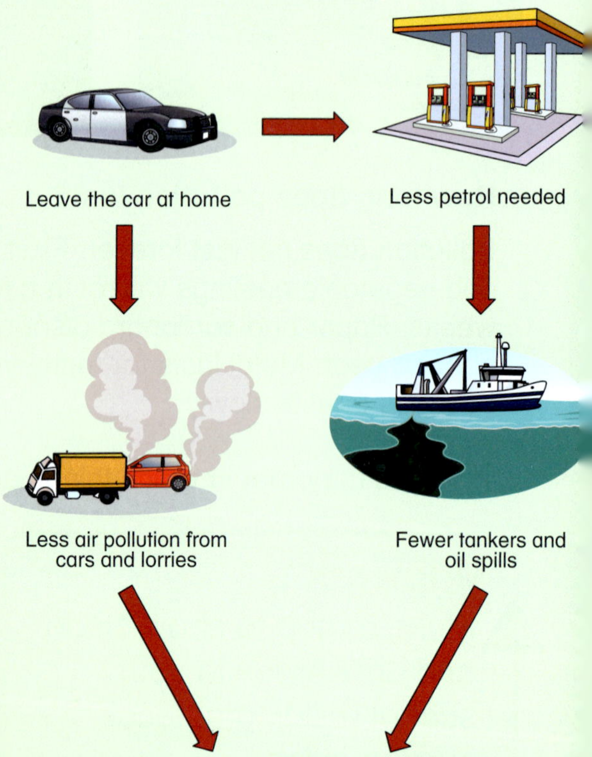
Leave the car at home
Less petrol needed
Less air pollution from cars and lorries
Fewer tankers and oil spills

Walking and cycling use only human energy. Cars and buses need fuel to run their engines and their exhust fumes pollute the air.

Cleaner environment

Discussion

- What does recycling mean?
- Why does it take time to change people's habits?
- How could you save energy yourself?

Investigation

- Devise a ten point 'Waste and Pollution' policy for your school.

Using 'green' goods

GREEN
CLEAN
wash

Washing powder and weed killer sometimes contain harmful chemicals. We can buy alternatives which do less damage to the environment.

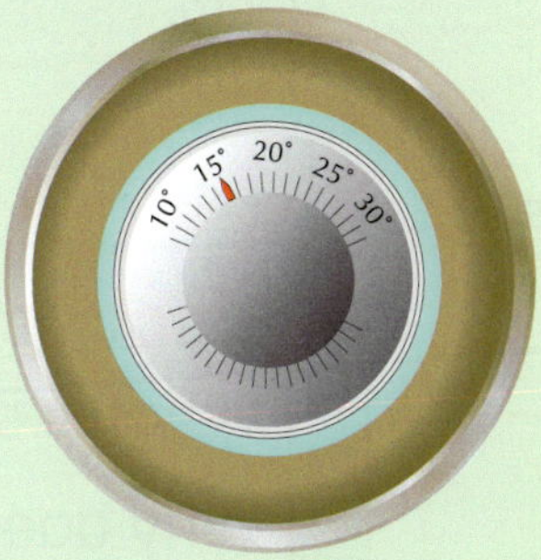
10°
15°
20°
25°
30°

Saving energy

If everyone switches off lights they do not need and truns down the heating thermostat in their homes, it would save a lot of energy.

Renewable energy

There are ways of making electricity which cause very little pollution. These make use of natural forces, like the heat of the sun and the power of the wind and water. They are called 'renewable' because, unlike coal, oil and gas, they will never run out.

▲ Wind farm.

Dams

Dams can trap water from rivers and seawater at high tide. When the water is released it is used to turn turbines.

Solar panels

Solar panels catch the energy from the sun and turn it into electricity.

Wind farms

In some exposed places, like marshes and the tops of hills, there are groups of windmills. When the wind blows, it turns the blades.

Mapwork

- Do a five-day survey of lights and machines which have been left on in your school. Show your findings on an outline map.

Lesson 3: Exploring clean energy

Key words

energy
fumes
power station
turbines

Can old power stations make clean energy?

▲ Drax power station has 12 cooling towers.

Drax power station in North Yorkshire, England is the largest coal-fired power station in western Europe. It supplies homes and factories throughout the country. Goods trains arrive at Drax throughout the day to deliver enough coal to keep the power station running.

The coal is burnt in boilers to make steam for the turbines. However, the coal also produces large amounts of smoke, ash and waste heat. At Drax they are trying to solve these problems and reduce the damage to the environment.

Parts of the power station have also been converted to burn wood pellets from North America. This makes a big difference to the amount of air pollution Drax causes. However, wood is much more bulky than coal and large areas of forest are needed to supply the trees.

Problem	Solution
Coal produces large quantities of fumes as it burns.	Limestone and water are sprayed onto the fumes to make them cleaner.
The coal produces thousands of tonnes of ash each year.	Some of the ash is sold to make plaster and cement.
The turbines produce a lot of steam.	Some of the heat from the steam is used in local greenhouses which grow tomatoes.

Data Bank

- Drax power station has the tallest factory chimney in the UK (259 metres).
- Drax can burn around 30,000 tonnes of coal a day.
- Drax generates 7% of UK electricity.

Discussion

- What is a power station?
- Why does Drax need to have such a tall chimney?
- What are the advantages and disadvantages of making electricity from coal?

A local investigation

At one primary school children in class 5 decided to investigate pollution problems in their local area. They recorded each problem with a circle on a survey sheet like the one opposite. They then added up the numbers they had circled. The score told the children how badly their local environment suffered from pollution. As they did the work the children also took photographs and made a map for a class display.

▲ Fumes from traffic.

▲ Noise from road works.

Mapwork

Draw a map or plan of the area you have studied in your pollution survey.

Pollution survey

Problem	No Problem	Some	A lot
Traffic exhaust	0	5	(10)
Factory fumes	0	5	(10)
Traffic noise	0	5	(10)
Aircraft noise	(0)	5	10
Polluted water	0	(5)	10
Litter	0	(5)	10
Overhead wires	0	(5)	10
Unpleasant smells	0	5	(10)
Factory noise	0	(5)	10
Vandalism/graffiti	(0)	5	10
Total		60	

0	25	50 or more
No pollution	Some pollution	A lot of pollution

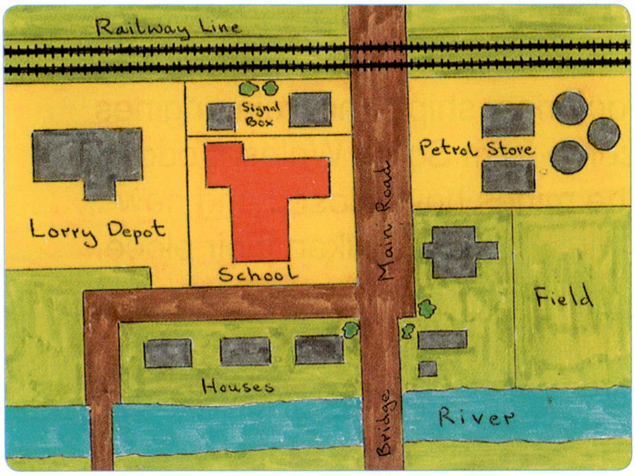
Railway Line
Signal Box
Petrol Store
Lorry Depot
School
Main Road
Field
Houses
Bridge
River

A map of the area around the school.

Investigation

Make a similar survey of pollution problems around your school.

Summary

In this unit you have learnt:

- about different forms of pollution
- what people are doing to solve pollution problems
- how to study pollution.

Unit 7 Wales

Lesson 1: Mountains and valleys

Wales

What is Wales like?

Wales is a country of hills, mountains, moors and valleys. It lies to the west of England and is about 200 kilometres long and 100 kilometres wide.

In the past there used to be a lot of mines in Wales. Roof slates for houses came from mines in the mountains of North Wales. Coal for factories, ships and train engines came from South Wales. Today the mines have closed and new industries have taken their place.

Two out of every three people in Wales live in the south of the country. In central Wales, people live in small towns, scattered villages and farms. In the north, the main settlements are on the coast where tourism is important.

Key words

Anglesey
Cambrian Mountains
Cardiff
River Wye
Snowdon
Swansea

Croeso i Gymru

▲ This sign says 'Welcome to Wales'. In some parts of Wales, Welsh is the main language.

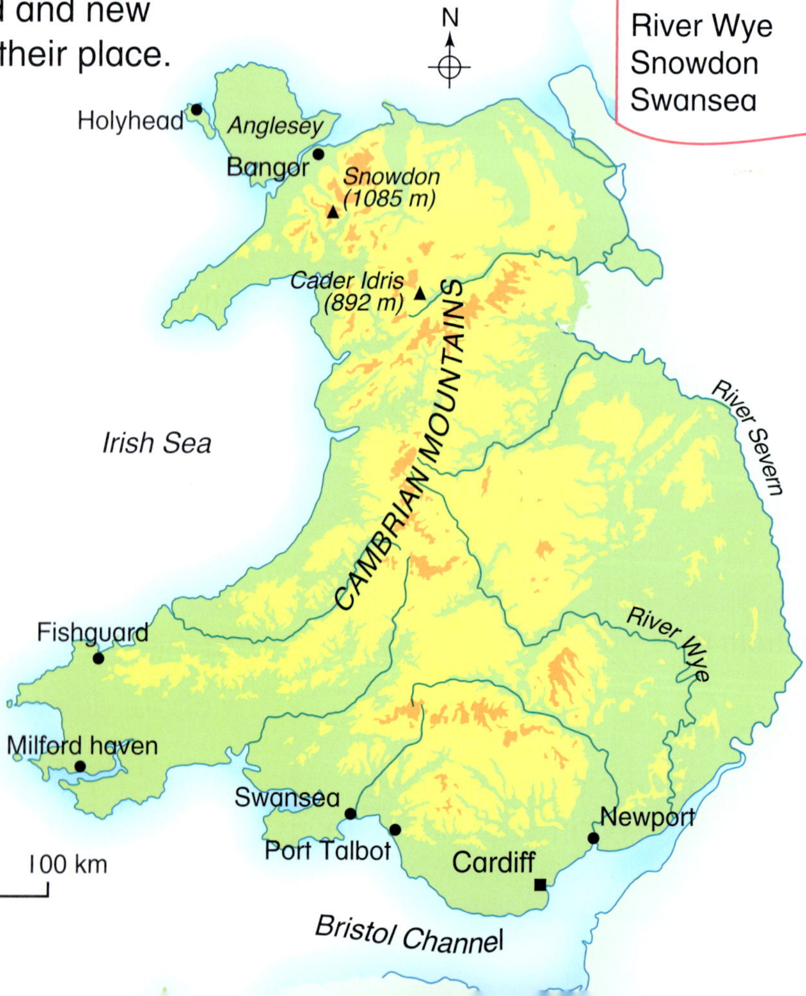
N
Holyhead
Anglesey
Bangor
Snowdon (1085 m)
Cader Idris (892 m)
CAMBRIAN MOUNTAINS
Irish Sea
River Severn
River Wye
Fishguard
Milford haven
Swansea
Port Talbot
Cardiff
Newport
Bristol Channel

Key

Over 500 metres
200-500 metres
0-200 metres

Scale

0 50 100 km

Discussion

- What are the main features of Wales?
- What used to come from Welsh mines?
- Why do you think most people in Wales live on the coast?

Rivers and landscape

- Snowdon and Cader Idris in the Cambrian Mountains are the highest peaks in Wales.
- The Wye is the longest river (209 km).
- Anglesey is an island off the northwest coast.

Transport

- The main airport is at Cardiff. The M4 motorway links Wales and England via the Severn Bridge.
- Ferries sail to Ireland from the ports of Fishguard and Holyhead.

Weather

- Most parts of Wales have a high rainfall.
- Strong winds affect the coast and other exposed places.
- On the highest mountains, snow lasts all winter.

▲ Hikers walking in the Snowdonia National Park.

Settlement

- Cardiff is the capital city.
- Swansea and Newport are important ports on the Bristol Channel.
- Bangor is a seaside resort on the north coast.

▲ A new stadium is part of the redevelopment of Cardiff Waterfront.

Work

- Most of the factories are in South Wales. Steel is made at Port Talbot.
- In the mountains, most farmers keep sheep. There are dairy farms in the lowlands and valleys.
- Forestry is important in hilly areas in North and Central Wales.

Mapwork

Using the map on page 38, work out the distance from Cardiff to Bangor (a) by land (b) by sea.

Investigation

Using the information from pages 38 and 39 create a fact file for (a) North Wales (b) South Wales.

Lesson 2: The story of Blaenavon

Key words

colliery
furnace
ironstone
lift shaft
heritage site

How is Wales changing?

▲ The view over Blaenavon and Big Pit today.

Blaenavon is an industrial town about 40 kilometres north of Cardiff, high in the mountains of South Wales. The name Blaenavon means 'the source of the river' in Welsh.

In 1789, during the Industrial Revolution, three furnaces were built at Blaenavon to make use of the local coal and ironstone. Within a few years the first coal mines had opened and the furnaces were producing thousands of tonnes of iron.

In 1852, in Victorian times, a railway was opened linking Blaenavon with Newport docks. The town grew larger as rows of stone cottages were built on the hillsides.

A large new colliery was also built. It was called Big Pit because it had a wide and deep lift shaft. The coal from Big Pit was excellent quality. It burnt with a great heat and left very little ash. This made it ideal for ships and railways engines all over the world.

▲ The ironworks in about 1800.

▲ Big Pit in 1900s when demand for coal was at its highest.

The 1920s and 1930s brought big changes. There was less demand for coal and many miners lost their jobs. The iron and steel works closed which made more people unemployed. However, the coal mine was modernised and continued working until 1980.

Today Blaenavon has a population of about 6000 people. There is a factory which makes parts for aeroplanes and a family business which makes cheese. In 1983, Big Pit opened as a coal-mining museum. It was made into a World Heritage Site in 2000 and has won several important awards since then.

BLAENAVON WORLD HERITAGE SITE
SAFLE TREFTADAETH Y BYD BLAENAFON
Former Mineral Railway Line
Blaenavon Ironworks
Big Pit
Blaenavon Heritage Town
UNESCO
©Torfaen County Borough Council Blaenavon World Heritage Site

▲ A tourist map of Blaenavon.

Discussion

- Why did Blaenavon become an industrial town?
- What was special about Big Pit?
- What is Blaenavon like today?

Mapwork

- List some of the differences between an Ordnance Survey map of Blaenavon and the map shown here.

Investigation

- Make a timeline starting in 1789 to show the history of Blaenavon.

Lesson 3: A visit to Big Pit

Key words

coal face
fan house
lift shaft
pit head
shaft
ventilation

What was it like to be a coal miner?

Glyn Davies used to work at Big Pit. He left school when he was 15. His job was to repair the steel ropes that pulled the lift up and down. Now he is a museum guide.

Glyn gives visitors a safety helmet with a cap lamp before taking them underground. He also gives each visitor a self-rescuer, just like the miners used to wear. Then everyone gets into the cage which quickly drops 90 metres to the bottom of the lift shaft.

It is very dark and musty in the mine. *"There are 40 kilometres of tunnels in Big Pit."* Glyn explains. *"They are held up with wooden props, bricks and steel arches. It was very important to get fresh air into the mine. A fan and ventilation doors helped to get rid of dangerous gases."*

Discussion

- How long did miners work in each shift?
- Why is it important to have museums like Big Pit?

▲ Visitors wearing their safety helmets.

Layout of underground workings

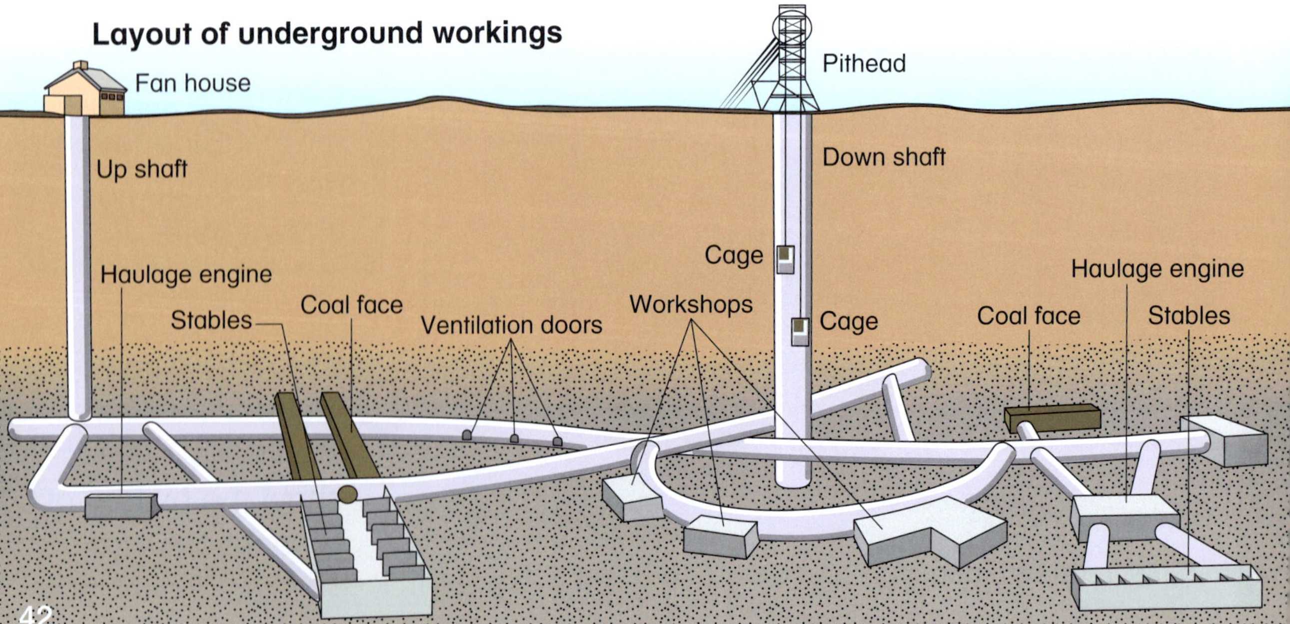
Fan house
Pithead
Up shaft
Down shaft
Cage
Haulage engine
Haulage engine
Stables
Coal face
Ventilation doors
Workshops
Cage
Coal face
Stables

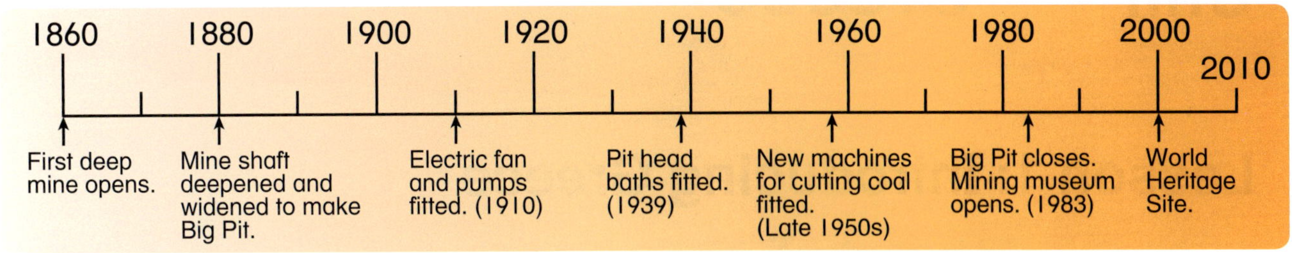
1860
1880
1900
1920
1940
1960
1980
2000
2010
First deep mine opens.
Mine shaft deepened and widened to make Big Pit.
Electric fan and pumps fitted. (1910)
Pit head baths fitted. (1939)
New machines for cutting coal fitted. (Late 1950s)
Big Pit closes. Mining museum opens. (1983)
World Heritage Site.

Glyn takes the visitors along the tunnels. He shows them the coal trucks and the rails they ran along. In some places, the ceiling is so low that everyone has to bend down. Water drips from the roof and the rocks are stained orange from the iron in the water. At one point, Glyn asks everyone to turn out their lamps. It is very dark. Then Glyn takes the visitors to the stables. In the past there were over 70 ponies at Big Pit. They lived in the dark all their lives and were used to pull the coal trucks.

Finally the visitors come to the coal face. Men worked in eight-hour shifts there. At night they cut away the coal. The morning shift loaded the trucks and the afternoon shift propped up the ceiling to make the new tunnel. It was a dangerous job and dust used to get into the miners' lungs which made them ill.

"As far as I am concerned," Glyn says, *"the coal from Big Pit was the finest in the world. Also we had very few accidents. The problem was that the coal seams ran out so the mine had to close. It was tough work but I am proud to have been a miner."*

Data Bank

- At one time over a thousand miners used to work at Big Pit.
- The coal face is around 100 metres below ground.
- Three million visitors have visited Big Pit since it became a museum in 1983.

Investigation

Find out about other World Heritage sites around the UK.

Mapwork

Make a 'spoke chart' with your school in the centre. Show museums, parks, visitor centres and other attractions children in your class have visited at the end of each spoke.

Summary

In this unit you have learnt:

- what makes Wales different from other countries in the United Kingdom
- about the workings of a coal mine
- how some parts of Wales are changing.

Unit 8 Greece

Lesson 1: Introducing Greece

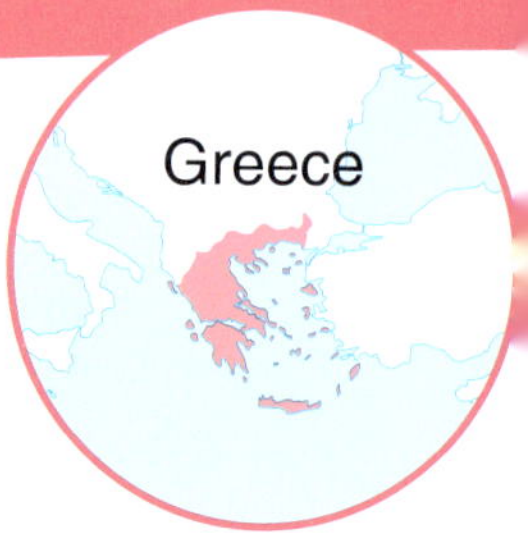
Greece

What is Greece like?

Three thousand years ago a great civilization flourished in Greece. The ancient Greeks made beautiful temples, statues and theatres. They also made important discoveries in science and mathematics. People have admired and copied their ideas ever since.

Today, Greece is part of the European Union. In addition to the mainland there are hundreds of islands scattered across the Aegean Sea. These are popular with tourists who come to enjoy the summer sun.

Key words

Aegean Sea	European Union
Athens	Mediterranean climate
Crete	Pindus mountains

Mapwork

Using an atlas work out how far it is from Athens to London and other European capitals.

BULGARIA
MACEDONIA
ALBANIA
Thessaloniki
PINDUS MOUNTAINS
Aegean Sea
GREECE
TURKEY
Ionian Sea
Patras
Piraeus
Athens
Corinth Canal
Amorgos
Crete
Key
Over 1000 metres
200-1000 metres
0-200 metres
Scale
0
200
400 km
N

Landscape

Much of Greece is covered by rocky mountains. These reach out into the sea in chains of islands.

▼ The Pindus Mountains.

Settlement

One in three people live in Athens, the capital city. Piraeus and Thessaloniki are the main ports.

Transport

Road and rail routes tend to follow the coast. Ferries link the islands with the mainland.

▼ The Corinth Canal links the Aegean and Ionian Seas.

Discussion

- What were the Greek people famous for in the past?
- Why is Greece such a popular place for summer holidays?
- In what ways is Greece different from the UK?

Climate

Greece has a Mediterranean climate. There is rain in winter and hot, dry summers.

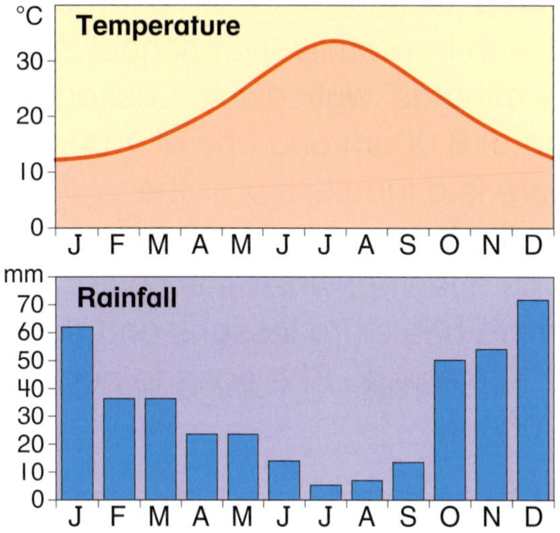
°C
Temperature
30
20
10
0
J F M A M J J A S O N D
mm
Rainfall
70
60
50
40
30
20
10
0
J F M A M J J A S O N D

Work

Factories make metals, chemicals, clothes and electronic goods. Farming and tourism are also important.

▼ Everyone in the family helps to harvest the olives.

Investigation

- Make up a fact file about Greece.

Lesson 2: Summer in Athens

Athens

What is the summer like in Athens?

Dimitra lives in Athens in a large flat with her family. Her school is five minutes' walk away. Lessons start at 8.00am and end at 1.00pm. There is a long break in the afternoon because it gets very hot. In the evening, when it is cooler Dimitra has extra lessons and does her homework. She goes to bed at 10.00pm.

▲ Dimitra and her mother on the ferry to Amorgos.

Sleep	School	Lunch and sleep	Extra lessons	Play or watch TV	Sleep

Midnight 8.00 am 1.00 pm 4.00 pm 6.00 pm 10.00 pm

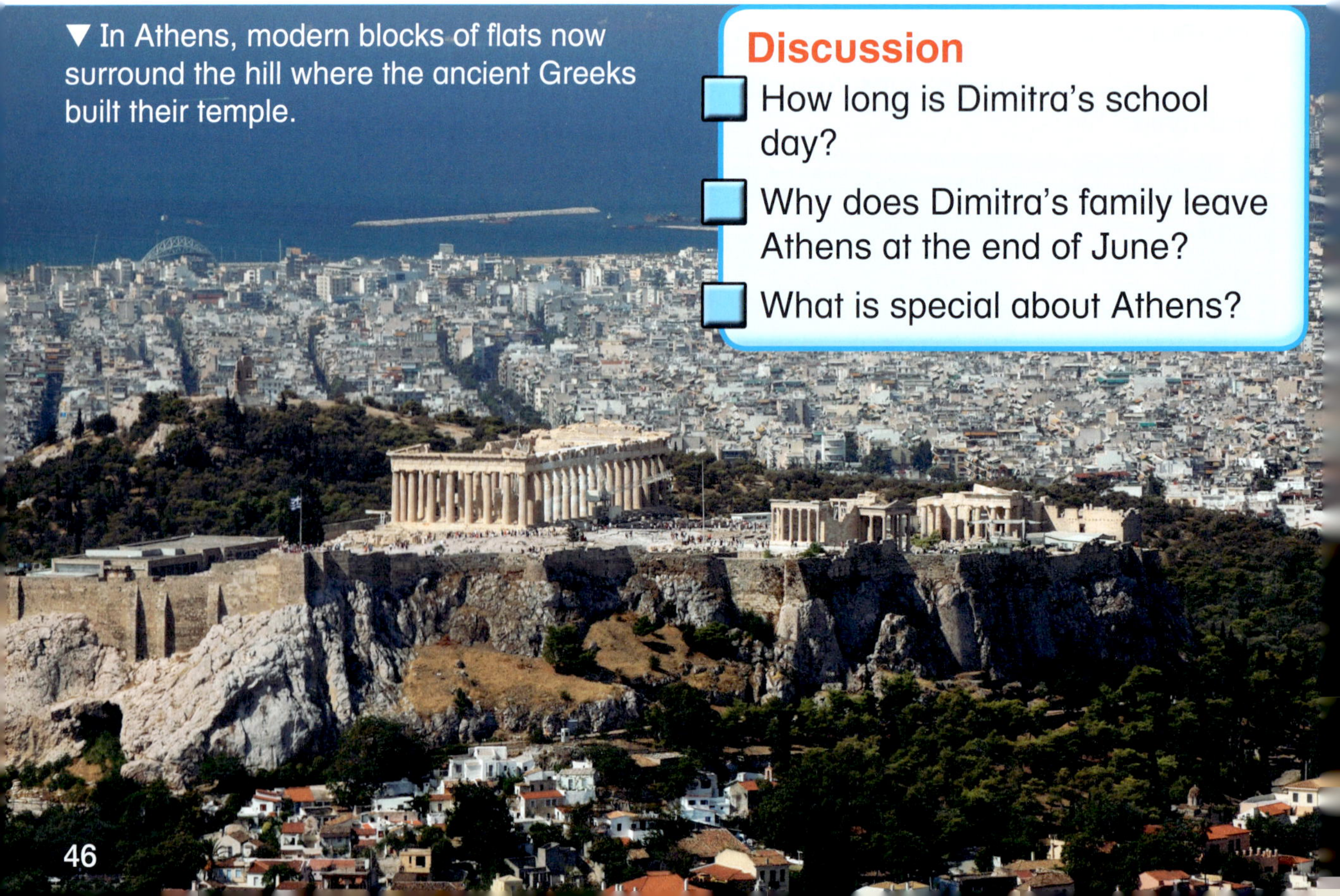

▼ In Athens, modern blocks of flats now surround the hill where the ancient Greeks built their temple.

Discussion

- How long is Dimitra's school day?
- Why does Dimitra's family leave Athens at the end of June?
- What is special about Athens?

Each year, at the end of June, Dimitra goes to stay with her cousins on the island of Amorgos. This is a good time to leave Athens because the heat and fumes from the traffic often cause smog. The Greeks call it nefos. Sometimes the nefos is so bad people are forbidden to use their cars.

Dimitra and her family travel by taxi to catch the ferry from Piraeus. At first, their journey takes them past office blocks and modern hotels. Then they go past the museum and the parliament building in Syntagma Square. The ruins of the Parthenon, an ancient temple, are in front them. Next they pass the old houses and narrow streets of the old town, or plaka. There are open-air markets and more ancient buildings.

When Dimitra last went to Amorgos the meltemi wind was blowing in the Aegean. *"I hope the sea won't be too rough for you today."* the taxi driver said.

Data Bank

- Around 800,000 people live in Athens.
- Athens holds the record for the highest temperature ever recorded in Europe (48°C).
- Over 16 million tourists visit Greece and Athens every year.

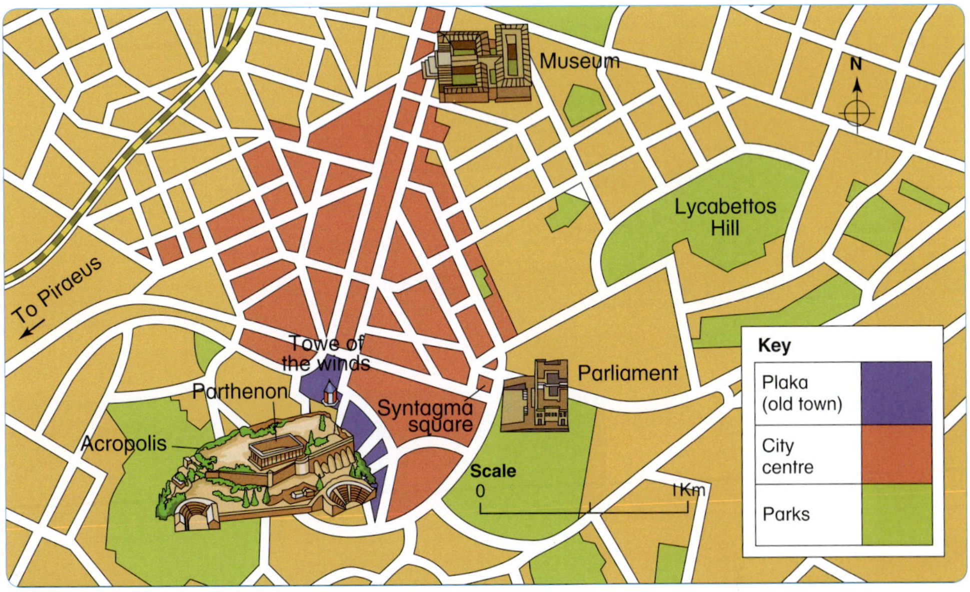
Museum
N
Lycabettos Hill
To Piraeus
Towe of the winds
Parthenon
Acropolis
Syntagma square
Parliament
Scale
0
1Km
Key
Plaka (old town)
City centre
Parks

Key words

Amorgos
market
meltemi wind
smog

Mapwork

Make drawings of the things which Dimitra sees on her journey through Athens. Put them in the correct order.

Investigation

Copy the timeline of Dimitra's day. Make a timeline of your own day underneath.

Lesson 3: A Greek island

Key words
beach
monastery
port
taverna

What is it like to visit Amorgos?

Dimitra and her family catch the ferry from Piraeus. *"It will take about 11 hours to reach Amorgos"*, her father tells her. *"We stop at four other islands before we get there."*

Amorgos is about 35 kilometres long and 6 kilometres wide. The countryside is rocky with steep hillsides. Drivers have to be careful on the narrow road that runs from the north to the south of the island.

Dimitra's cousins live in a fishing port called Egiali. It has a large harbour and a wide sandy beach. The town is built on a hill and has narrow streets and alleyways with steep steps.

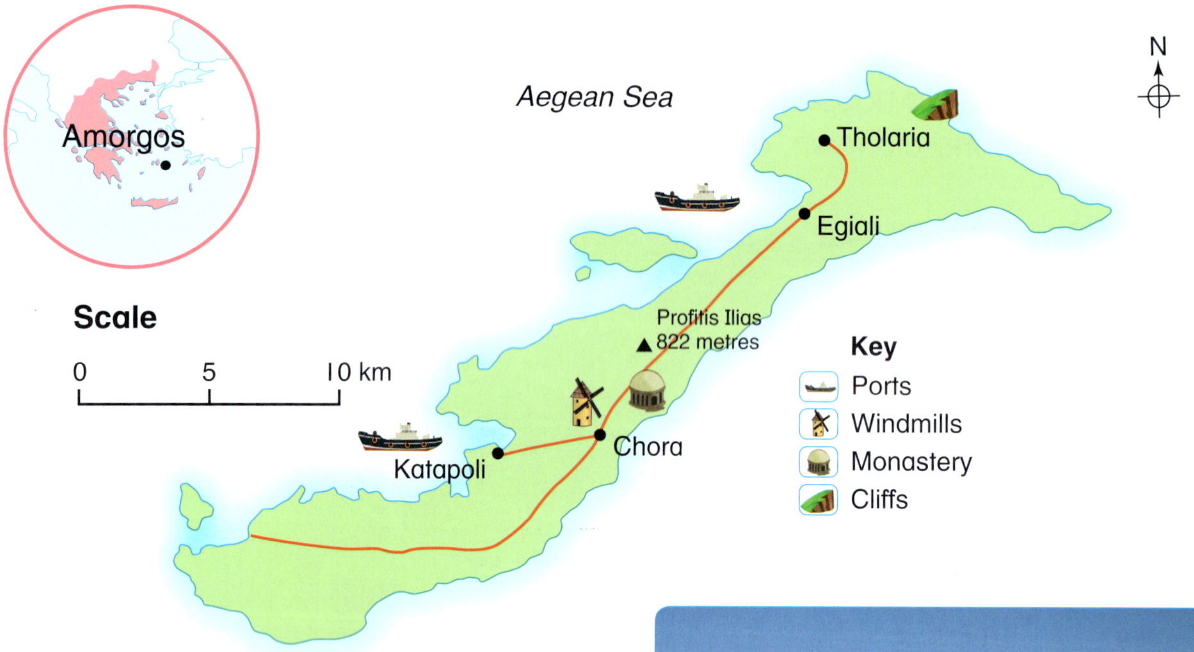
Amorgos
Aegean Sea
N
Tholaria
Egiali
Profitis Ilias
822 metres
Chora
Katapoli
Scale
0
5
10 km
Key
Ports
Windmills
Monastery
Cliffs

Data Bank

- Ferries link Piraeus to 60 islands.
- There are usually 250 sunny days a year in the Greek islands.
- There are more than 200 islands in Greece – the largest is Crete.

▲ Most of the ferries which go to the islands leave from Piraeus harbour.

▲ Most of the houses in the town are whitewashed to keep them cool in summer.

▲ Dimitra's collection of things which remind her of the holidays.

▲ Lunch at the taverna.

Discussion

- How many islands are linked by ferry to Pireaeus?
- What is Egiali like?
- What would you most like to do if you were staying at the taverna on Amorgos?

Mapwork

Draw a route map for tourists travelling from Tholaria to Katapoli by bus. Describe what they might see along the way.

Investigation

Write down three things you think Dimitra might want to tell you about Athens and Amorgos.

Summary

In this unit you have learnt:

- about the environment of Greece
- what Athens is like in the summer
- about the Greek islands.

Unit 9 North America

Lesson 1: Introducing the Caribbean

Caribbean

What is the Caribbean like?

The Caribbean consists of the Caribbean Sea, its islands and surrounding coasts. There are over 700 islands and reefs in the Caribbean and it is famous for its tropical beaches.

The Caribbean is also known as the West Indies. Some of the islands such as Jamaica and Trinidad were once British colonies. They are now members of the Commonwealth and play sports with the UK. Others used to belong to France or Spain.

Key words

Caribbean Sea	Trinidad
Cuba	Tropic of Cancer
Jamaica	Tropic of Capricorn

Discussion

- How many islands are there in the Caribbean?
- How is the Caribbean linked to Europe?
- What do you think you might like or dislike about the Caribbean?

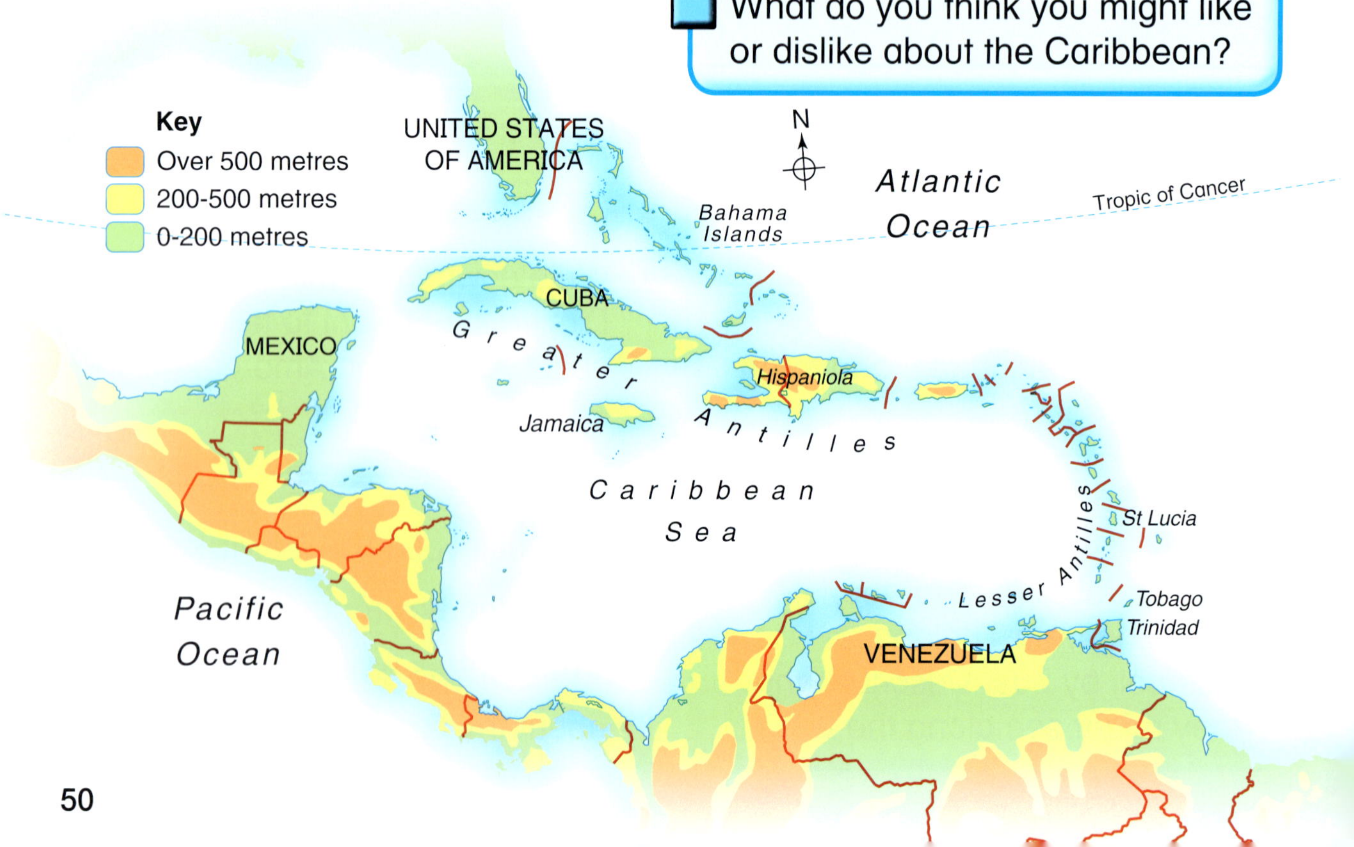
Key
Over 500 metres
200-500 metres
0-200 metres
UNITED STATES OF AMERICA
N
Atlantic Ocean
Tropic of Cancer
Bahama Islands
CUBA
MEXICO
Greater Antilles
Hispaniola
Jamaica
Caribbean Sea
St Lucia
Lesser Antilles
Tobago
Trinidad
Pacific Ocean
VENEZUELA

Landscape

There are two main chains of islands - the Greater and Lesser Antilles.

▼ The Pitons in St Lucia were created by volcanoes.

Climate

The Caribbean has a tropical climate. This means that there is a wet season and a dry season and that it is warm throughout the year.

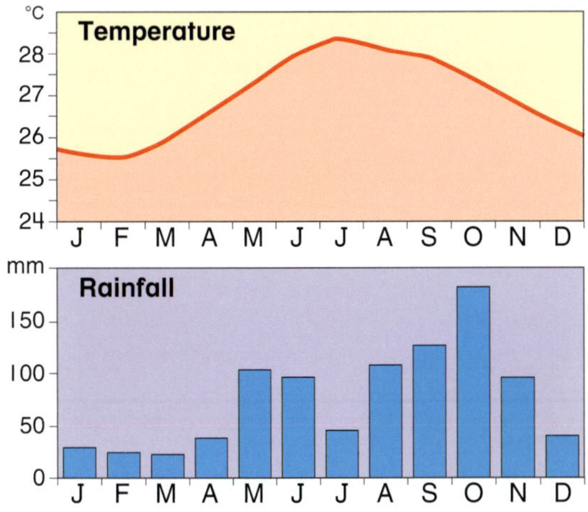
°C
Temperature
28
27
26
25
24
J F M A M J J A S O N D
mm
Rainfall
150
100
50
0
J F M A M J J A S O N D

Mapwork

Working from an atlas, make a list of islands which are on or near (a) the Tropic of Cancer (b) the Tropic of Capricorn.

Culture

Most people speak either a European language or a creole (local dialect). The Caribbean is famous for rap, reggae and other types of music.

Environment

Large areas of rainforest have been cleared for farming. Around the coast mangrove swamps and coral reefs have been damaged.

▼ From July to November there is a danger of hurricanes.

Tourism

Tourism provides jobs for many Caribbean islands. Cruise ships bring visitors from the USA and Europe. Some people have private yachts.

Investigation

Find out about hurricanes, how they form and the routes that they take.

Jamaica

Lesson 2: Finding out about Jamaica

What is Jamaica like?

Jamaica is an island in the Caribbean Sea. It is about half the size of Wales. Around the coast there are sandy beaches and palm trees. The highest point is Blue Mountain Peak which is 2258 metres high.

Jamaica is much hotter than the United Kingdom. There are only two seasons. From November until April it is fairly dry, from May to October it is much wetter. The mountains protect some places from the rain. Sugar, coffee and tobacco are important crops in Jamaica. The country also has valuable supplies of bauxite – the raw material for aluminium.

Jamaica
Caribbean Sea

▲ Satellite image of Jamaica.

Mapwork

Make a map and write notes about one other Caribbean island of your choice for a class display.

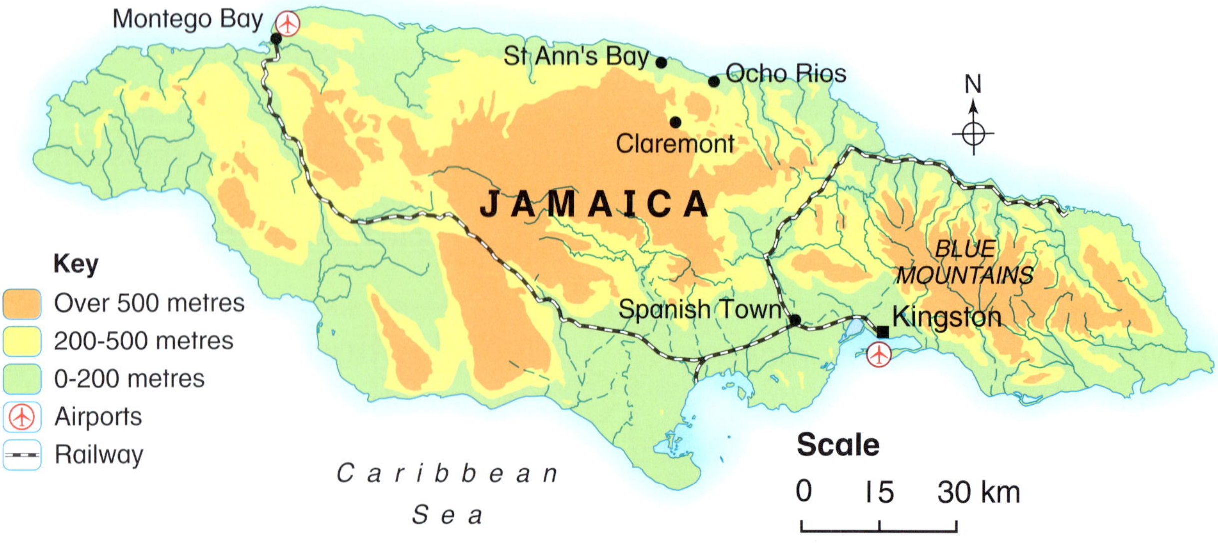
Montego Bay
St Ann's Bay
Ocho Rios
Claremont
JAMAICA
N
BLUE MOUNTAINS
Spanish Town
Kingston
Key
Over 500 metres
200-500 metres
0-200 metres
Airports
Railway
Caribbean Sea
Scale
0 15 30 km

Lesson 3: Living in Jamaica

How is Jamaica changing?

Eunice Grant is a teacher. She was brought up in Claremont, Jamaica but has lived in England for many years. The children from Riverside Primary School asked Eunice to talk about her childhood.

"We lived in a bungalow with a verandah in the front. At the back there was an orchard where we kept chickens, cows and goats.

In the fields all around us people grew sugar cane, bananas, oranges and lots of other fruits. They also grew vegetables like yams, sweet potatoes and peppers.

▲ A field of sugar cane.

▲ There are trees all around the bungalow.

School started at 7.30 in the morning so we could have our lessons before it got too hot. I wore a tunic and white blouse for school. The boys wore khaki trousers and white shirts."

Investigation

Write a few sentences about some of the different ways Jamaica is linked to the UK through trade, sport, music and history.

Sunday in Jamaica

"Sunday was the best day of the week in Jamaica. In the morning people put on their smartest clothes and went to church. The women and girls all wore hats and the men put on their suits.

After the church service we had dinner. We had rice, beans, chicken and salad and a cool fruit drink with ice cream. When I was a child I used to like this better than anything else.

After dinner people played cricket or sat around talking and listening to music. Sometimes we all went to the beach at St Anne's Bay. When I was young I used to look for new shells for my collection."

Key words

bungalow	habitats
palm trees	mangrove
sugar cane	parrot
sweet potatoes	seashore
verandah	

▲ Women singing in the church choir.

Discussion

- What made Sunday special for Eunice?
- What food did Eunice's family get from the land around the bungalow?

▲ Sunday afternoon on the beach.

Jamaica today

"Claremont has changed quite a bit since I was young. There are many more houses now. Also people have more possessions like mobile phones and computers.

Tourists from Europe and North America come to Jamaica for their holidays. They fly to the airport at Kingston or Montego Bay and stay at hotels along the coast. The tourists bring money but the seashore is being damaged. Mangrove forests are cleared away to make room for new buildings. Parrots and other animals which live in the forests are losing their habitats.

▲ Mangrove trees have roots which come out of the water at low tide.

When I first came to England everything seemed very different. I found the weather very cold. English people stay indoors much more. In Jamaica we are always going outside and meeting our friends. I have got used to it now."

▲ Today Yellow-naped Amazon parrots are very rare in Jamaica.

In the towns many people work in clothes factories. Some people cannot find work and are very poor. Others have plenty of money.

Investigation

Make a fact file of six things you have learnt about Jamaica.

Summary

In this unit you have learnt:

- about the landscape of Central and North America
- about everyday life in Jamaica
- about how people can tell you about a place.

Unit 10 Africa

Lesson 1: Introducing Africa

What is Africa like?

Africa is the second largest continent. It lies to the south of Europe across the Mediterranean Sea. The Sahara Desert covers most of North Africa. Further south there are grasslands, rainforests and mountains. In Africa grasslands are called savannahs.

Over the centuries there have been great nations in Africa. These included Benin and Great Zimbabwe. In Victorian times the Europeans divided Africa between themselves to build empires.

Today there are nearly 60 countries in Africa. Most of them became independent in the 1960s and 1970s. Today, Africa supplies the rest of the world with food and metals at very low prices. Cotton, coffee, tea and groundnuts are some of the crops which are grown for export. Copper, gold, diamonds and precious stones come from southern Africa. Elsewhere there are valuable supplies of oil.

▲ Snow-capped Mount Kilimanjaro rises above the savannah in Kenya.

▲ Cape Town and Table Mountain at the tip of southern Africa.

Key words

Cairo	River Nile
Lake Victoria	Sahara Desert
Mount Kilimanjaro	

Discussion

- What four main types of landscape are found in Africa?
- What happened to Africa in Victorian times?
- What goods does Africa supply to the rest of the world?

Africa

Atlas Mountains
Cairo
Sahara Desert
Atlantic Ocean
River Nile
River Niger
Key
Mountain
Desert
Grassland
Forest
Lagos
Equator
River Congo
Lake Victoria
N
Mt Kilimanjaro
Indian Ocean
Scale
0
1000
2000 km
River Zambezi
MADAGASCAR
Kalahari Desert
Johannesburg
Cape Town

Data Bank

- The Sahara Desert is the largest desert in the world (about 8.5 million sq km in area).
- The Nile, Congo, Niger and Zambezi are the main rivers in Africa.
- Lake Victoria is the largest lake.

Mapwork

Using an atlas name a country in each of the four main landscape types.

Investigation

Make up ten questions for a quiz about Africa.

▲ There are a lot of gold mines around Johannesburg.

Lesson 2: Kenya

Kenya

Finding out about Kenya

The landscape of Kenya is one of the most varied in Africa. There are deserts, farmlands, savannahs, rainforests and extinct volcanoes. The coast has beautiful beaches fringed with palm trees.

Most people in Kenya live in the central highlands where the cool climate is good for farming. Coffee, tea, maize, fruit and vegetables are important crops.

Kenya was once ruled by Britain but became independent in 1964. English and Swahili are the main languages. The capital, Nairobi, is growing fast.

▲ Tourists come to Kenya to visit the game parks.

▲ There are motorways and office blocks in the centre of Nairobi.

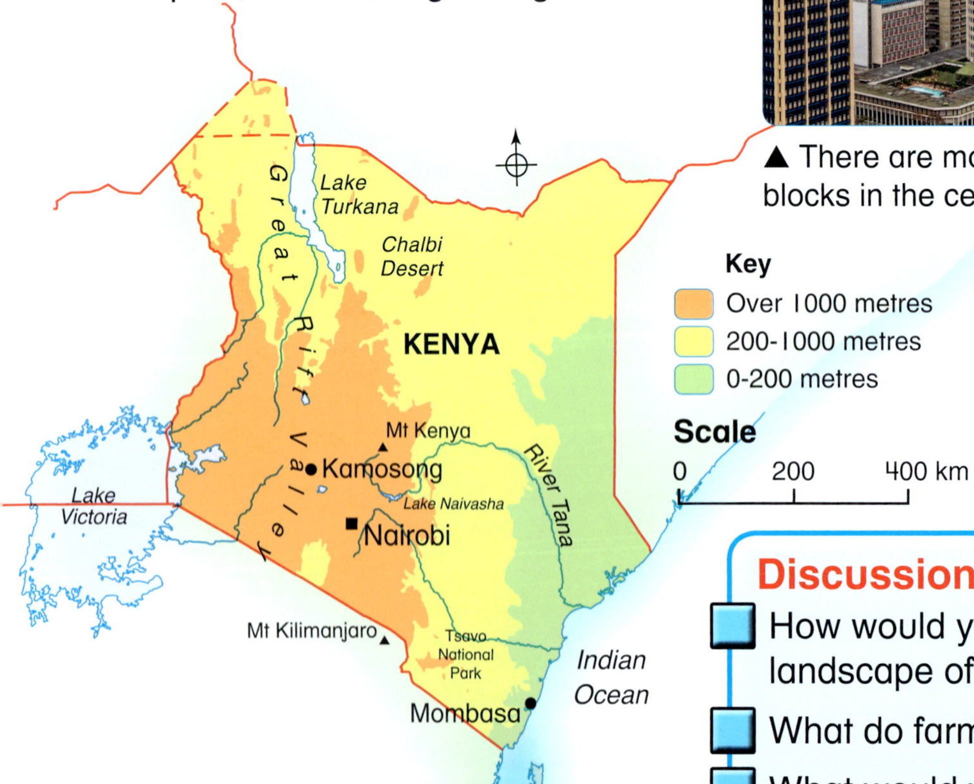
Lake Turkana
Great Rift Valley
Chalbi Desert
KENYA
Mt Kenya
Kamosong
Lake Naivasha
River Tana
Nairobi
Lake Victoria
Mt Kilimanjaro
Tsavo National Park
Indian Ocean
Mombasa
Key
Over 1000 metres
200-1000 metres
0-200 metres
Scale
0
200
400 km

▲ Kenya is more than twice the size of the United Kingdom but it has half the population.

Discussion

- How would you describe the landscape of Kenya?
- What do farms produce in Kenya?
- What would you find different if you lived with Miriam's family?

Making links with a school in Kenya

At St Peter's School pupils wanted to find out more about Kenya. Their class teacher set up a link with a school in a village called Kamosong about 300 kilometres from Nairobi. In their first letters the children told each other about the place where they lived. They also drew maps of their local area.

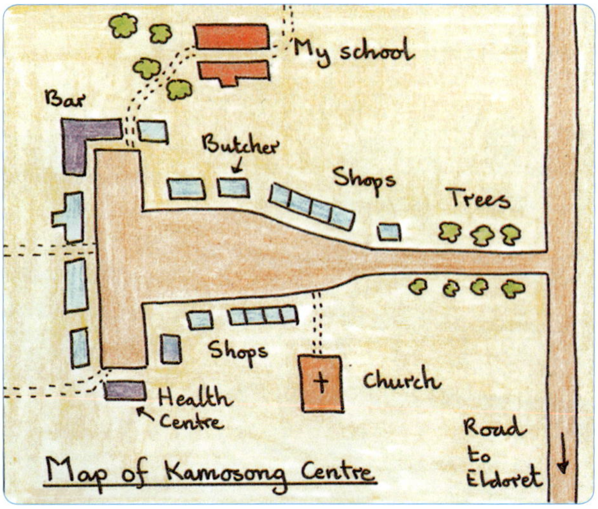
My school
Bar
Butcher
Shops
Trees
Shops
Church
Health Centre
Road to Eldoret
Map of Kamosong Centre

▲ Miriam lives about five kilometres from Kamosong Centre.

Kamosong Primary School
PO Box 2755
Eldoret
KENYA

Dear Holly,

I am very thankful for receiving your letter on the third of this month and having known more about your family and school activities.

My hobbies are drawing, making friends and taking photos. I have four brothers and three sisters plus myself. This makes a total of eight children in our family.

I get up at 6.00am. My favourite lessons are art, Swahili and science. When I am not at school I go to the garden to weed my pyrethrum. I get £15 a month from the garden. I read novels in my free time. I would also like to come one day to England.

Our house is made of earth and thatched with grass. I am among the girl guide group and we always practice once a week.

Tell me if you are a boy or a girl. I am a girl aged 12.

Hoping to hear from you next time.

Yours,

Miriam

Key words

central highlands
extinct volcanoes
Kamosong
pyrethrum
savannah
Swahili

Investigation

Write a description of Miriam's life using these headings:
(a) family
(b) hobbies
(c) school day.

A parcel from Kenya

Investigation

Select six items from the display table. Write a sentence saying what each one tells you about Kenya.

KENYA
STANDARD
MATATUS FARES RAISED BY 100 p.c.
Florence the blue flamingo
KENYA
ROUTE MAP

1. A jar of coffee.
2. A shaker made from bundles of reeds.
3. A strip of cloth with traditional patterns.
4. A newspaper with headlines about a local rise in taxi fares.
5. Kenyan bank notes.
6. A sandal bought in the local market made from a recycled car tyre.
7. Green beans grown in Kenya and sold in shops in the UK.
8. Photographs of the school and houses where the children live.
9. Postcards of animals in one of the safari parks.
10. Water pots.
11. A children's book.
12. A map showing places to visit around Kamosong.

One day the children received a parcel from their friends in Kamosong. It contained a selection of newspapers and postcards, a few bus tickets, a strip of cloth and some photographs which the children had taken of each other. There were also some interesting stamps on the outside of the parcel.

The children arranged everything on a display table. Some of the class added fresh fruit and tins of food from Kenya which they had bought in the local shops. Their teacher added a few Kenyan ornaments and some musical instruments.

As the children continued with their project, they found other sources of information. They read books about Kenya and watched a television programme.

Discussion

- What is the most unusual object in the display?
- What would you like to investigate about Kenya?
- Which six things would you put in a box to describe your own life?

Lesson 3: Living in Kenya

How is Kenya changing?

Here are some of the things which the children at St Peter's found out about life in Kenya today.

Key words

annotated map
drought
game park
shanty town

Weather

In recent years the rainfall has been unreliable. Heavy rain has caused floods while other regions have suffered droughts.

▼ Droughts are causing the Chalbi Desert to spread into surrounding areas.

Cities

Nairobi, Mombasa and other cities are growing larger. People are moving there in search of work.

Game parks

There are over 15 game parks in Kenya. These are popular with tourists but local people are sometimes forced out to make more space for animals.

Farming

Some people earn money selling beans and roses to Europe. These plants take valuable water which is needed to grow crops for local people.

▼ Aeroplanes bring roses from Lake Naivasha for sale in the UK.

Mapwork

Draw a map of Kenya. Annotate it with notes round the edge saying how Kenya is changing.

Investigation

Design an advertisement to attract tourists to a Kenyan game park. Include information about its location, the weather and the journey from the UK.

Summary

In this unit you have learnt:

- about the cities and countryside of Africa
- how links with another school can help provide information
- about some of the issues in Kenya today.

Glossary

Bulk carrier
A lorry designed to carry heavy goods like sand and gravel.

Bungalow
A type of house, originally from India, often one storey high with a verandah.

City
A very large settlement which in the UK usually has a cathedral.

Channel
The route taken by water in a stream or river.

Climate
The pattern of weather over many years.

Compass
An instrument with a suspended magnetic needle that points to the North and South Poles.

Crop
Plants which farmers grow and harvest such as wheat, apples or bananas.

Engineer
A person who repairs mechanical or electrical devices.

Extinct volcano
A volcano which was once active but which is now completely dead.

Fish stocks
The quantity of fish in the world.

Fumes
Harmful gases which damage the health of people, plants and animals.

Harbour
A sheltered area on the coast for boats and shipping.

Hurricane
Violent storms which form in tropical oceans and which do great damage when they reach land.

Life boat
A special boat for rescuing sailors and other people from storms at sea.

Mangrove forest
forests which grow along the seashore, which can survive in salty water and cope with the changes brought by the tide.

Marine reserve
An area of sea or ocean where fish, sea creatures and plants are protected by law.

Minerals
Materials which are found naturally in rocks and soils. Examples include coal, oil, gold and silver.

Nuclear waste
Radioactive material which is left from industrial activity e.g. making electricity.

Oil platform
A large structure for drilling wells and extracting oil at sea.

Ranging pole
Red and white striped pole for measuring slopes and changes in level.

Resource
Something which people find useful.

Settlement
The places where people live, such as villages, towns and cities.

Skyscraper
A very tall building, often built round a steel frame and used as flats or offices.

Suburbs
The outer edges of a town or city where there are often houses and schools.

Tributary
A stream or river that flows into a larger river.

Tropics
Parts of the world where the sun is directly overhead at least once a year.

Turbine
A machine with blades that generates power.

Verandah
A covered area along the edge of a house or bungalow which is open to the weather.

Index

Primary Geography Book 5
Collins
An imprint of HarperCollins Publishers
Westerhill Road
Bishopbriggs
Glasgow
G64 2QT

ISBN 978-0-00-756361-6

Imp 007

British Library Cataloguing in Publication Data
A catalogue record for this book is available from the British Library.

Printed and bound by Printing Express, Hong Kong

Acknowledgements

Additional original input by Terry Jewson

Cover designs Steve Evans illustration and design

Illustrations by Jouve Pvt Ltd pp 2, 3, 7, 11, 12, 21, 26, 28, 29, 33, 34, 35, 42, 47

This product uses map data licensed from Ordnance Survey® with the permission of the Controller of Her Majesty's Stationery Office.
© Crown copyright. Licence number 100018598

Adapted extract on p17 from *Plain Tales from the Raj* by Charles Allen, published by Futura Paperbacks © 1975 by Charles Allen and the BBC

Hand drawn map of Blaenavon on p41 reproduced by permission of © Torfaen County Borough Council Blaenavon World Heritage Site

Photo credits:

(t = top b = bottom l = left r = right c = centre)

© Capt' Gorgeous/Flickr.com p42; © Chrysaora/Flickr.com p54b; © geographyalltheway.com/Flickr.com p13t; © johnbraid/Shutterstock.com p19tr; © Mark Turner/Flickr.com p40t; © NASA Goddard Photo and Video/Flickr.com p32; © National Museum Wales p40bl, p40br; © Peter Q/Flickr.com p54tr; © Stephen Scoffham p49tr, p60

All other images from www.shutterstock.com

Contents

1 It's your planet!

geog.1 pages 4-5

The photo below shows planet Earth. Viewing Earth from space allows you to see the Earth as a whole. It's an amazing sight!

1 Write down three words that best describe what Earth looks like from space.

____

____

____

2 Imagine you are flying around Earth in a spaceship. Put these three words into a paragraph that best describes what Earth looks like from space.

____

____

____

3 Identify the five places, A to E, shown on the satellite photo of the earth below.

A ______________________

B ______________________

C ______________________

D ______________________

E ______________________

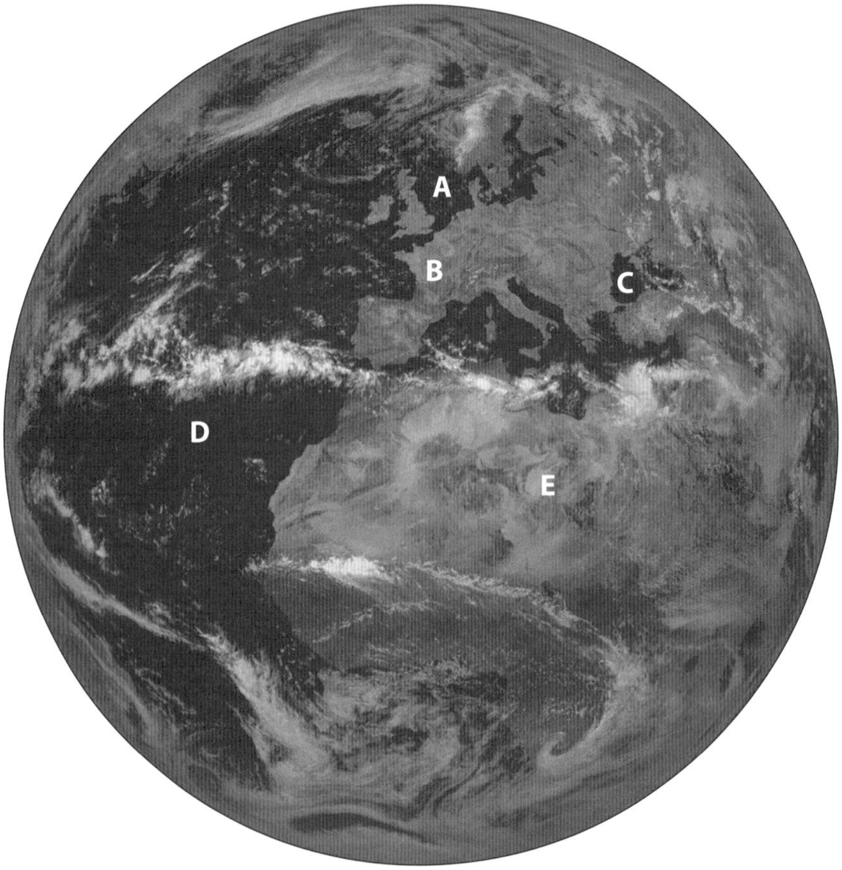
A
B
C
D
E

1.1 Earth's story: it begins with a bang

This is about how it all started.

geog.1
pages 6-7

1 Read the five sentences below. Circle T or F at the end to say whether they are true or false. Use pages 6-7 in the pupil book to help you.

a The Big Bang happened about 13 million years ago. T / F

b The first star was a hot glowing ball of hydrogen. T / F

c The Sun was formed in a galaxy called the Milky Way. T / F

d The Earth is around 4.5 billion years old. T / F

e The atmosphere around Earth had very little water vapour in it. T / F

BANG

2 The paragraph below describes what the Earth was like when it formed. Draw a picture in the space below that shows what you think the Earth looked like. Be as creative as you can!

> The Earth looked very different to today. It would have looked red and orange, not the blue, green and white of today. The surface would have been covered with hot molten rock, the size of many oceans. There was no life, and no water. The atmosphere contained no oxygen, but a mix of deadly poisons.

1.2 Earth's story: life develops

From simple cells to humans – this is about the story of evolution.

4th edition
geog.1
pages 8-9

1 Look at these statements. They are all about how fossils are formed but are jumbled up. Write a number from 1-6 in the box to put them in the correct order.

Reptile dies ☐

Ocean floor is pushed upwards, becoming land ☐

Reptile living in ocean ☐

Reptile body falls to ocean floor ☐

Fossil found by human ☐

Reptile buried under mud and sand ☐

2 Imagine you are Freddie, a reptile living near the bottom of the ocean waiting to become a fossil. Draw a series of pictures in the spaces below to show how Freddie Fossil was formed and eventually found by a human.

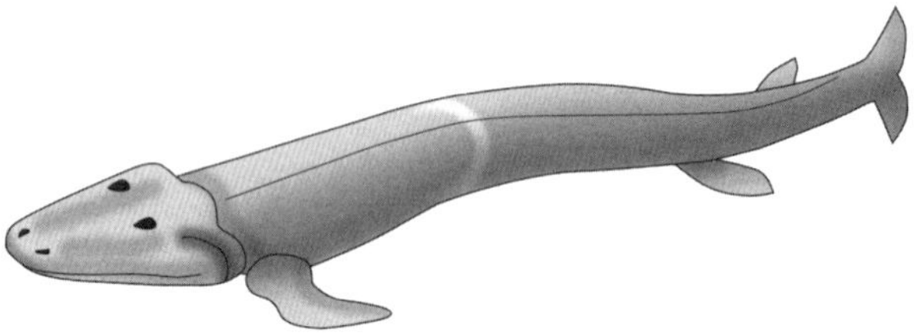

Tip: Remember to put your drawings in the correct order. **!**

1.3 Earth's story: the time scale

Here you will learn more about the Jurassic period.

geog.1
pages 10-11

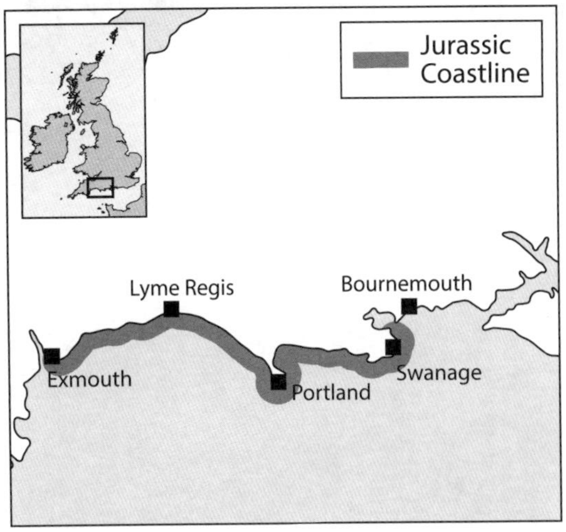
Jurassic Coastline
Lyme Regis
Bournemouth
Exmouth
Portland
Swanage

The map shows the 95 mile long Jurassic Coast of Dorset and East Devon. It is one of the most spectacular coastlines in England, and is so important that it is now a World Heritage site - just like the Great Wall of China! It has classic cliffs, adorable arches and spectacular sea stacks. This part of the coast is the only place on Earth where we can study 185 million years of geological time. It is the finest record of the Triassic, Jurassic and Cretaceous periods in the world.

1 Become a detective and do some research about the Jurassic coast. Find out some amazing facts about its arches, its cliffs and the fossils you can find. Write your answers by the side of the arch, cliff and fossil shown below.

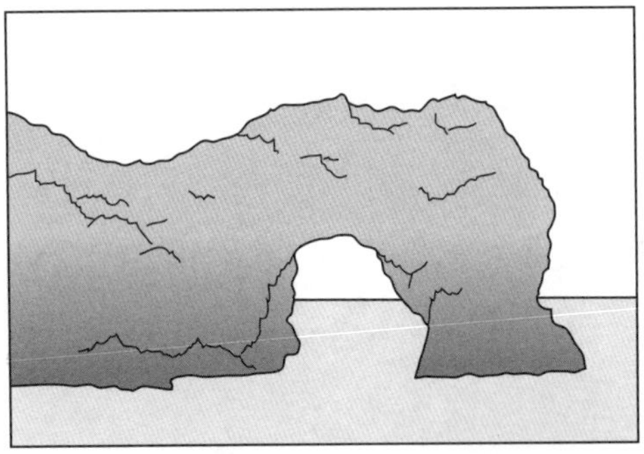

____

____

____

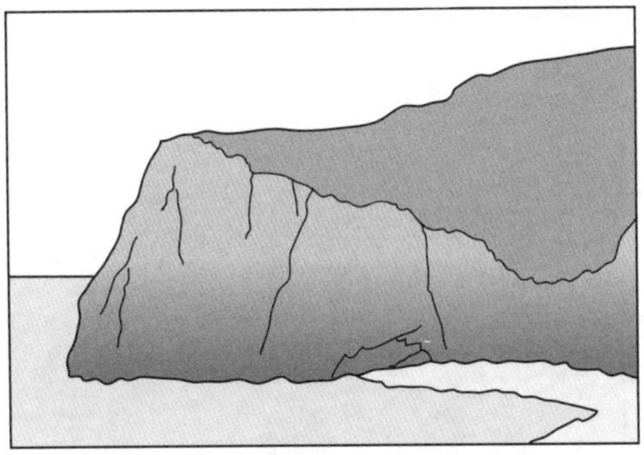

____

____

____

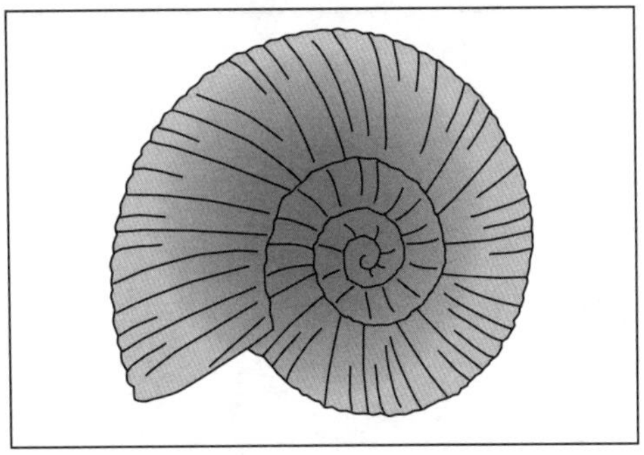

____

____

____

1.4 Our time on Earth

geog.1
pages 12-13

This is about humans appearing on Earth.

1 Early human movement can be called the Journey of Man. Fill in the gaps in these paragraphs. Choose words from the box.

Humans first appeared in East ____________. It took us ____________ years to reach Britain. However, as we moved 60 000 years ago ice was a problem. Many places were covered in ice ____________. The Ice Age was cold so humans left Britain and went to warmer parts of ____________. When the ice began to melt, we came back!

Some ocean floor got ____________ of water which meant that we could walk across areas that used to be sea. Land created like this is often called a land ____________. When humans left Africa we had a dark skin, but as we moved away from the ____________ the sunlight became weaker and our skin ____________ changed. Now humans have a range of skin colours across the world!

Equator	drained	Europe	bridge
Africa	colour	20 000	sheets

2 Early man did not have many possessions when making these journeys. Think about possessions or items that might have helped him, what would he have needed?

Label the drawing with your ideas, giving reasons.

Item:

Reason:

Item:

Reason:

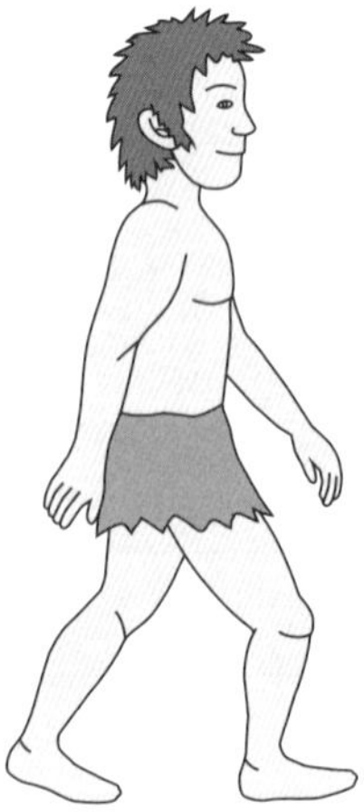

Item:

Reason:

Item:

Reason:

1.5 Our place on Earth

This is about where you would like to live.

4th edition
geog.1
pages 14-15

1 Everybody can think of things that would make a place great to live in. These are often called the *features* of a place. In the box below circle five features that would be important in your ideal place.

countryside

leisure centre

railway

shops

park

bus station

roads

industry

cinema

houses

2 In the space below make a sketch of your ideal place to live using the five features that you have chosen.

3 Why have you planned your place like you have? What reasons can you think of to explain? Give as many as you can.

1.6 Earth: a very special planet

This is about where you live – planet Earth!

pages 16-17

1 The earth spins as it travels non-stop around the sun. Why do you not fall off?

____

____

2 Planet Earth is full of life. Fill in the gaps in the following passage using words from the box below.

Elephants are the ____________ land animals on earth. There are only around 600 000 of them left: they are an ____________ species. New forms of ____________ are being found all of the time. This new life is found on the land and also in the ____________.

oceans	endangered	largest	life	animal

Did you know, for every one of me, there are 11 166 of you humans out there!

3 Choose one fact about planet Earth. Draw a diagram in the box below to show your chosen fact.

1.7 Changing Earth

This is about how natural forces and humans change our planet.

geog.1 pages 18-19

1 a Each of the photos shows how planet Earth is changing. Under each photo write a caption saying why. Choose from the captions given below.

> Water in the river scrapes and shapes the land as it flows
>
> Villages, towns and cities use a lot of land
>
> Hot rock can change what the landscape looks like
>
> Factories use valuable resources and sometimes cause pollution

b When you have written your captions, underline the natural reasons in one colour, and the human causes in another.

2 Humans are destroying the Earth. Do you agree?

1.8 It's all geography!

This is about how being nosy could make you a good geographer!

geog.1 pages 20-21

1 Geography can be divided into three different strands, *physical*, *human* and *environmental*. Explain what each one means in your own words.

2 a Now write down any topics you can think of that are part of geography (try to think of at least eight).

b Circle any topics that are *physical* geography in one colour, *human* geography in another, and environmental geography in a third.

c Do any of your topics include all three strands? Circle them in a fourth colour.

3 a Have a look at this photo. Underneath, brainstorm questions you could ask about it (can you think of six?).

b Are you able to answer any of your questions? How would you find answers to the others?

2 Maps and mapping

geog.1
pages 22-23

Stick a map, or part of a map, in this box.
It can be any sort of map – just as long as it's a map.
Then answer the questions below.

1 What does your map show?

____

2 What's the best thing, or most interesting thing, about your map?

____

3 How would you define a map? Finish this sentence:

A map is ______________________________________

____

4 Write down when and how, or why, you last used a map.

____

____

2.1 Mapping connections

This is about how we are connected to people and places all over the world – and how this can be shown using maps.

geog.1

pages 24-25

Like Walter, you are connected to hundreds of places. Some of them are in Europe.

0
300
600
900 km
N
ICELAND
NORWAY
SWEDEN
FINLAND
ESTONIA
RUSSIA
LATVIA
LITHUANIA
DENMARK
IRELAND
NETHERLANDS
BELARUS
UNITED KINGDOM
POLAND
UKRAINE
GERMANY
CZECH REPUBLIC
BELGIUM
SLOVAKIA
SWITZERLAND
MOLDOVA
AUSTRIA
HUNGARY
FRANCE
ROMANIA
PORTUGAL
CROATIA
SLOVENIA
SERBIA
BOSNIA-HERZEGOVINA
BULGARIA
ITALY
KOSOVO
MONTENEGRO
TURKEY
SPAIN
MACEDONIA
ALBANIA
GREECE
TUNISIA
MOROCCO
ALGERIA

a Think of five countries in Europe that you have a connection with. List these in the first column of the table.

b In the second column, add the reason for your connection.

c In the table, shade each country a different colour.

d Now shade in the countries on the map that you are connected to. Shade them so that they match your table.

Country	*Connection*

2.2 A plan of Walter's room

This is about what plans are, and what the scale of the plan tells you.

geog.1

pages 26-27

1 A drawing of something seen from above is called a **plan**.

Match the drawing of a chair with the correct plan (tick the correct one).

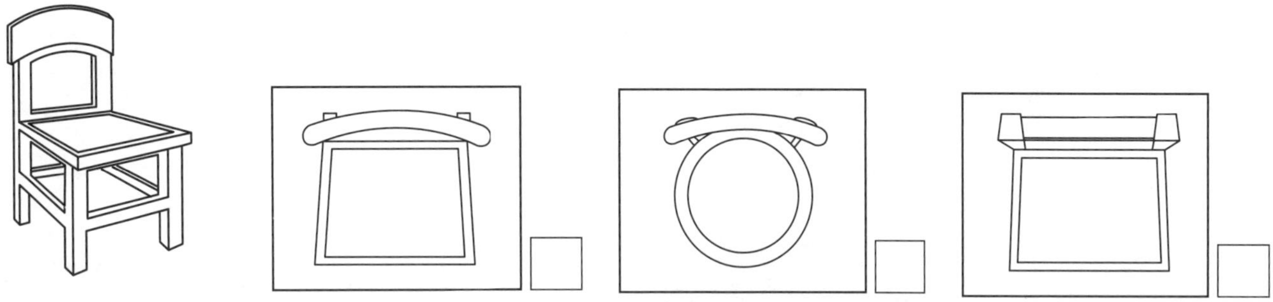

2 Now look at this plan of a bedroom. 1 cm on the plan represents 40 cm in the room. That is the **scale** of the plan.

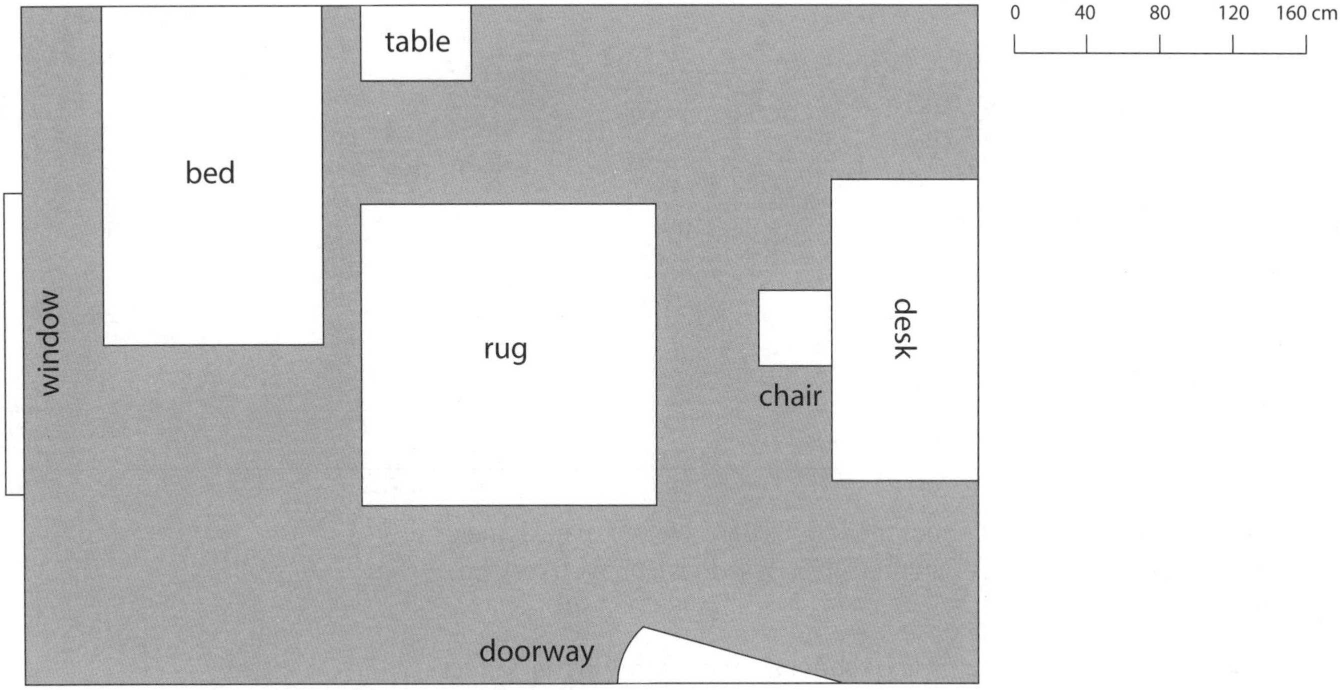

table
bed
window
rug
chair
desk
doorway
0
40
80
120
160 cm

a On the plan, the window is this wide: ____________________

So, in real life the window is _____ cm wide. (Fill in the gap.)

b Now measure the length of the bed and fill in the gaps below.

On the plan, the bed is _____ cm long.

This means it is _____ cm long in real life.

c Something in the room is 60 cm wide in real life. What is it? ____________________

d What in the room is 160 cm x 80 cm in real life? ____________________

2.3 Your mental maps

This is about your very own, personal mental maps.

geog.1 pages 28-29

1 **a** Think about a place you know well – it may be your local area or a park for example. In the space below draw a mental map of that area.

b Add labels to show your feelings about the various parts of your map. Some of your feelings may be happiness, excitement, fear or sadness, but you may be able to think of others as well.

Think of a symbol for each of your feelings and draw the symbol in the correct place on your map.

c Show your map to a partner. Write down the thing that they like best about your map.

2.4 Real maps

This is about how maps are built up.

geog.1
pages 30-31

This photo shows a railway bridge over the River Tamar in Devon.
Your task is to draw a sketch map of the same place.
Don't forget a key!

Key

2.5 Using grid references

This is about finding places on a map, using grid references.

4th edition
geog.1
pages 32-33

1 Fill in the gaps in this sentence, choosing from the words in the box.

A good map has five things: a __________, a frame around it, an arrow to show __________, a __________ and a __________.

title	note
north	scale
east	lock
key	river

2 Look at this map.

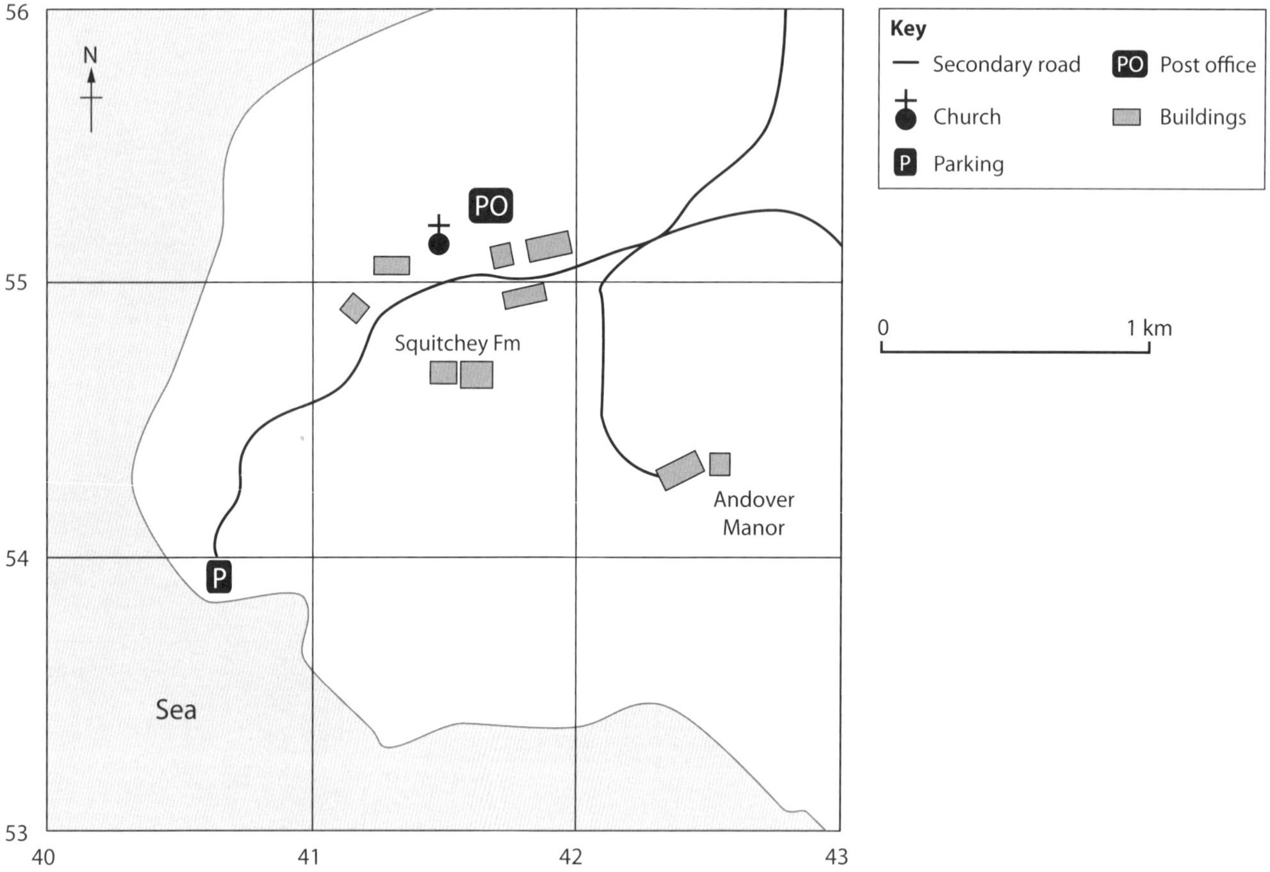

56
55
54
53
40
41
42
43
N
PO
Squitchey Fm
Andover
Manor
P
Sea
Key
Secondary road
Church
Parking
PO Post office
Buildings
0
1 km

Give a four-figure grid reference for:

a Squitchey Farm __________ **b** Andover manor __________ **c** the church __________

3 What is at this grid reference on the map?

a 407539 __________ **b** 414552 __________ **c** 416553 __________

4 Now add two more things to the map. Name them, and give their six-figure grid reference.

a __________ is at __________

b __________ is at __________

2.6 How far?

4th edition
geog.1
pages 34-35

This is about how to find the distance between two places on a map.

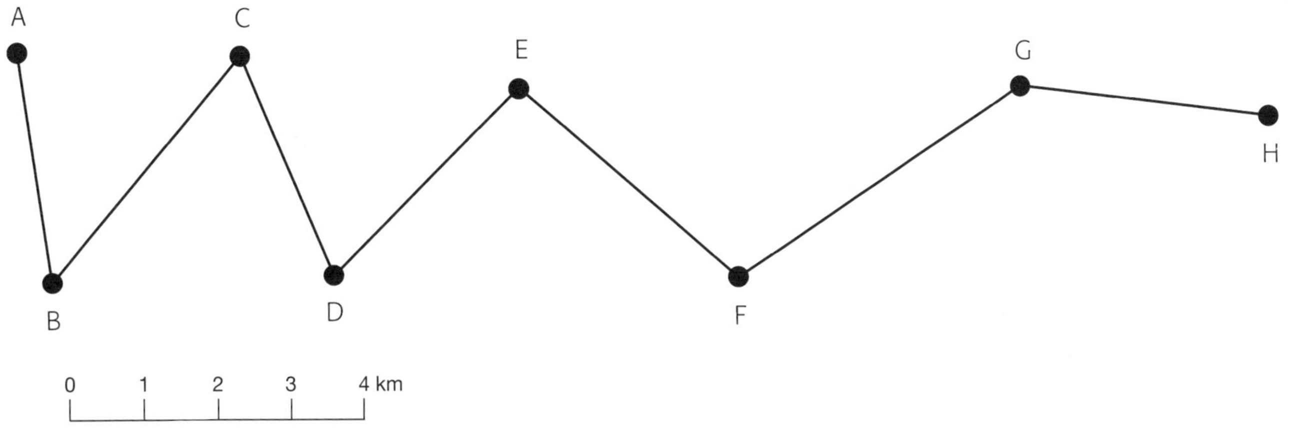
A
C
E
G
H
B
D
F
0
1
2
3
4 km

1 How far is it as the crow flies:

a from A to C? ____________________ **b** from A to H? ____________________

2 How far is it by road:

a from A to C? ____________________ **b** from A to H? ____________________

3 Have a look at the map on page 21.

a How far is it as the crow flies from Hella Point (in square 3721) to Pordenack Point (square 3424)? ____________________

b How far is it by road between St Buryan (4125) and Trethewey (3823)?

____

4 a Follow these instructions.

Drive east from Land's End for just over a kilometre. Take a right turning and follow the road for 1.8 km. Turn left and follow the short track to the end.

Where do you end up? ____________________

b Now give instructions (as in c) to someone who wants to travel from Treen (3922) to Trebehor (3724).

____

____

____

____

2.7 Which direction?

This is about how to give and follow directions, using N, S, E and W.

geog.1

pages 36-37

1 Draw an arrow going in these directions (**a** has been done for you).

a to the south **b** to the north east **c** to the west **d** to the south west

e to the north west **f** to the east **g** to the north **h** to the south east

2 Look at this grid. Now follow the instructions on the right.

6						
5						
4						
3						
2						
1						
	A	B	C	D	E	F

- Start in square E5.
- Go three squares W. Colour this square *red*.
- Go two squares S. Colour this square *green*.
- Go two squares SE. Colour this square *blue*.
- Go one square NE. Colour this square *yellow*.
- Go four squares NW. Colour this square *purple*.
- Go four squares E and one square S.
  What square do you end up in?

____

3 Name the direction the arrow is coming **from**.

a 

from the ______________

b 

from the ______________

c 

from the ______________

d 

from the ______________

2.8 Ordnance Survey maps

geog.1
pages 38-39

This is about what OS maps are, and what they show, and how to use them.

This OS map shows part of the Land's End peninsula in Cornwall.

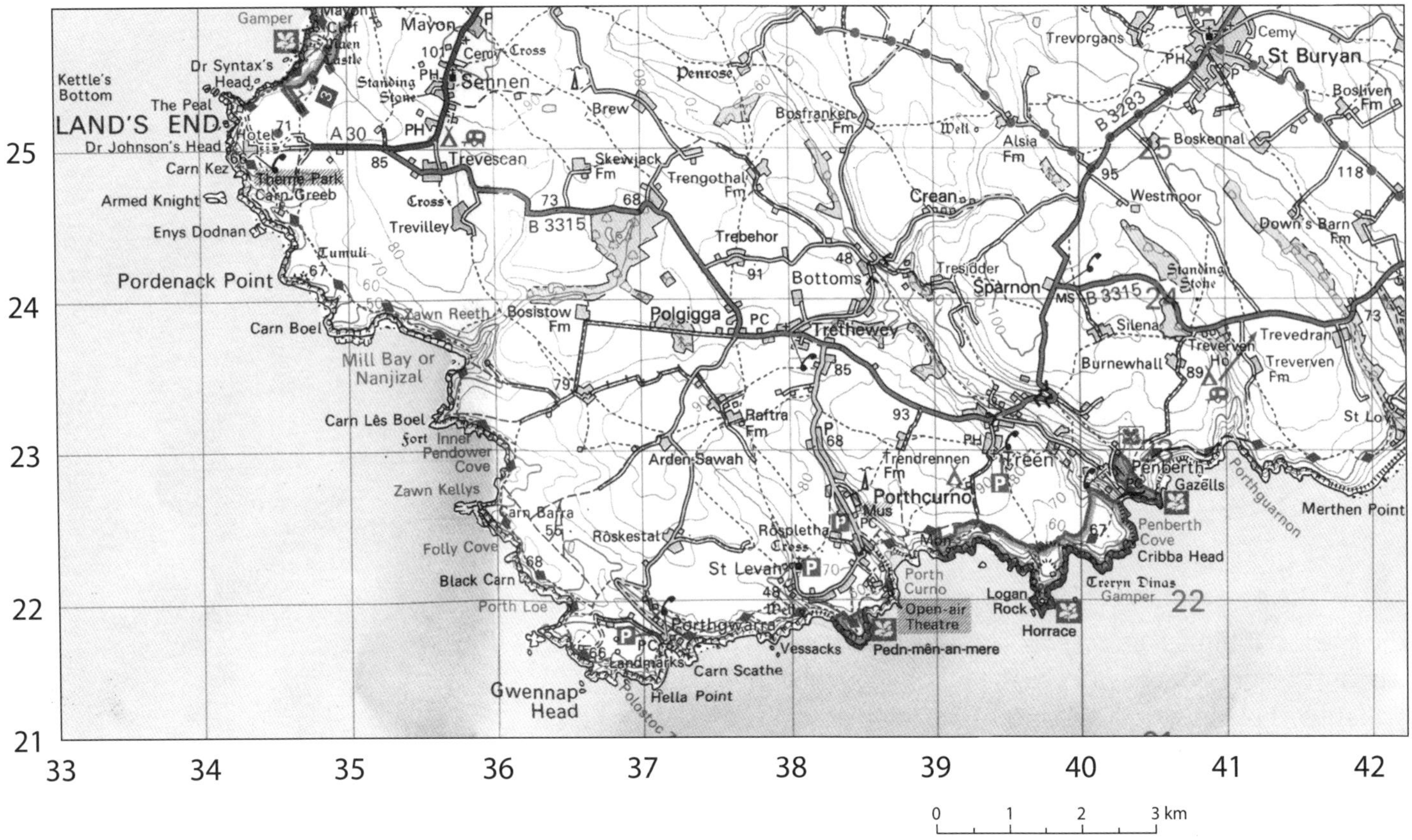
LAND'S END
Dr Johnson's Head
Carn Kez
Armed Knight
Enys Dodnan
Pordenack Point
Carn Boel
Mill Bay or Nanjizal
Carn Lês Boel
Zawn Kellys
Folly Cove
Black Carn
Porth Loe
Gwennap Head
Hella Point
Carn Scathe
Kettle's Bottom
Dr Syntax's Head
The Peal
Gamper
Mayon
Sennen
Standing Stone
Trevescan
Theme Park
Carn Greeb
Tumuli
Zawn Reeth
Bosistow Fm
Trevilley
B 3315
A 30
Brew
Skewjack Fm
Trengothal Fm
Penrose
Bosfranken Fm
Trebehor
Bottoms
Polgigga
Trethewey
Raftra Fm
Arden-Sawah
Roskestal
St Levan
Rospletha Cross
Porthgwarra
Vessacks
Porthcurno
Trendrennen Fm
Treen
Porth Curno
Open-air Theatre
Pedn-mên-an-mere
Logan Rock
Horrace
Treryn Dinas
Penberth Cove
Cribba Head
Gazells
Merthen Point
Porthguarnon
St Loy
Treverven Fm
Trevedran
Burnewhall
Silena
Sparnon
Tresidder
Crean
Alsia Fm
Westmoor
Boskennal
Down's Barn Fm
Standing Stone
B 3283
St Buryan
Trevorgans
Bosliven Fm
Well
25
24
23
22
21
33
34
35
36
37
38
39
40
41
42
0
1
2
3 km

1 What is at this grid reference?

a 387219 ______________________

b 385253 (Hint: Fm means farm) ______________________

c 366249 ______________________

d 345242 ______________________

2 Find one of these on the map and give a six-figure grid reference for it.

a a car park ______________________

b a church ______________________

c a public phone ______________________

d a camp site ______________________

3 What clues are there on the map that the Land's End peninsula gets lots of visitors? Give as many as you can.

____

____

____

2.9 How high?

This is about how height is shown on an OS map.

geog.1

pages 40-41

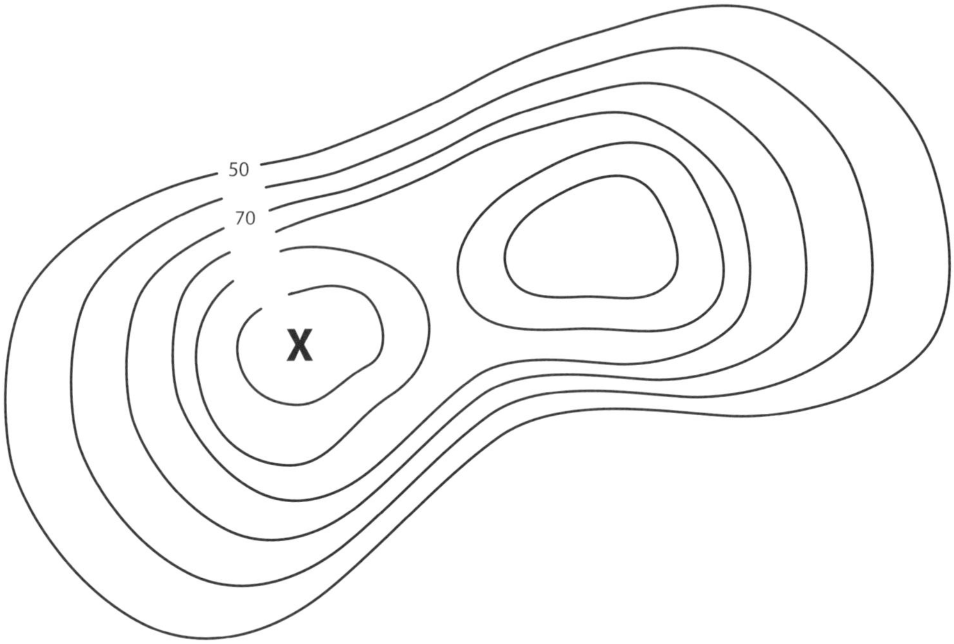
50
70
X

1 The lines on this map are contour lines. Everything along a contour line is the same height above sea level. The number on the line shows the height in metres. These contour lines are at 10 m intervals.

a Write in the four missing labels.

b Above what height is the land at X? ______________________________

c Colour in all the land above 80 metres.

d Write a label to show where the slope is steep.

e Now write a label to show where the slope is gentle.

2 Now look at the OS map on page 21. About how high above sea level is:

a Trevilley (358246)? ______________________________

b Raftra Farm (376233)? ______________________________

3 **a** Complete this sentence:

Another way that OS maps show how high a place is by using ______________

______________________________. These give the exact height at a spot, in metres above sea level.

b Can you find an example of this on the OS map on page 21? Give a four-figure grid reference.

____

2.10 Where on Earth are you?

This is about the special grid lines we use to say where places are on Earth.

4th edition
geog.1
pages 42-43

1 Cross out the wrong word in these sentences.

- **a** The lines that circle the Earth from top to bottom are lines of longitude / latitude.
- **b** The lines that circle the Earth from side to side are lines of longitude / latitude.
- **c** The 0° line of latitude is called the Equator / Arctic Circle.
- **d** The 0° line of longitude is called the Equator / Prime Meridian.

2 Write these statements as coordinates:

- **a** 33° north of the Equator, 20° east of the Prime Meridian. ______________________
- **b** 44° south of the Equator, 40° west of the Prime Meridian. ______________________

3 Look at the map.

- **a** Finish labelling the lines of latitude.
- **b** Label the five main lines of latitude: the Equator, the Tropic of Cancer, the Tropic of Capricorn, the Arctic Circle, and the Antarctic Circle.
- **c** Now finish labelling the lines of longitude.
- **d** Label the Prime Meridian.
- **e** Shade the area between the Tropic of Cancer and the Tropic of Capricorn in a 'warm' colour. Label this region 'The tropics'.
- **f** Shade the Arctic and Antarctica in a 'cool' colour. Label these regions.

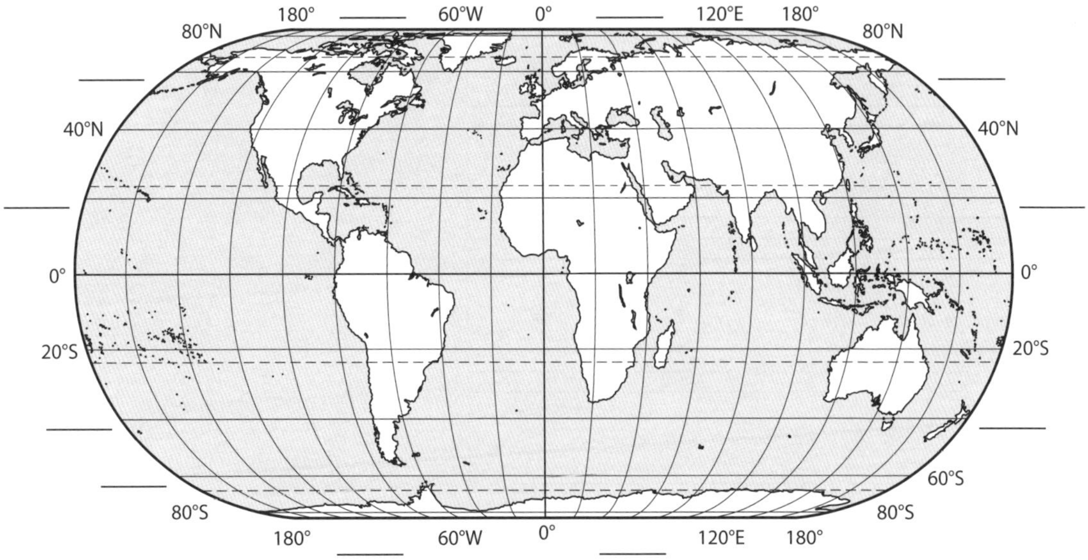
180°
60°W
0°
120°E
180°
80°N
80°N
40°N
40°N
0°
0°
20°S
20°S
60°S
80°S
80°S
180°
60°W
0°
120°E
180°

4 Mark three dots on the map, each one in a different continent.
Label them X, Y, and Z. Write the coordinates of your places here:

X ____________________ Y ____________________ Z ____________________

3 What about the UK?

geog.1
pages 44-45

1 Here are traditional flowers or plants associated with the countries of the UK. Can you say which part of the UK they stand for?

_______________ _______________

_______________ _______________

2 What is your image of the UK? Write down the words or phrases that come to mind when you think of the UK under the headings below.

Sport

____

Music

____

Achievements/Inventions

____

Famous people

____

Food

____

3.1 Your island home

This is about the forces that shaped the British Isles – and about Britain's main physical features.

4th edition
geog.1
pages 46-47

1 This paragraph explains how our island home became an island. Choose words from the box to fill in the gaps.

Once upon a time, the British Isles lay at the ____________, as part of a giant ____________. When this broke up, they drifted ____________ as part of Europe. As they drifted, over millions of years, they went through many ____________. They became desert. They were frozen in ____________. They were drowned by the ____________. They had earthquakes and eruptions. They got pushed and squeezed until ____________ grew. And then they got ____________ from the rest of Europe.

ice	cut off	changes	sea
crust	continent	mountains	currents
north	equator		

2 Here are some features of the British Isles, but they're all jumbled up. Unscramble the words and then use them to label the map below.

vrier sernve
rierv htmaes
eisglhn nanchel
rthno aes
isirh sae
verri nttre
alke drictist
ninespen
rthno estw landshigh

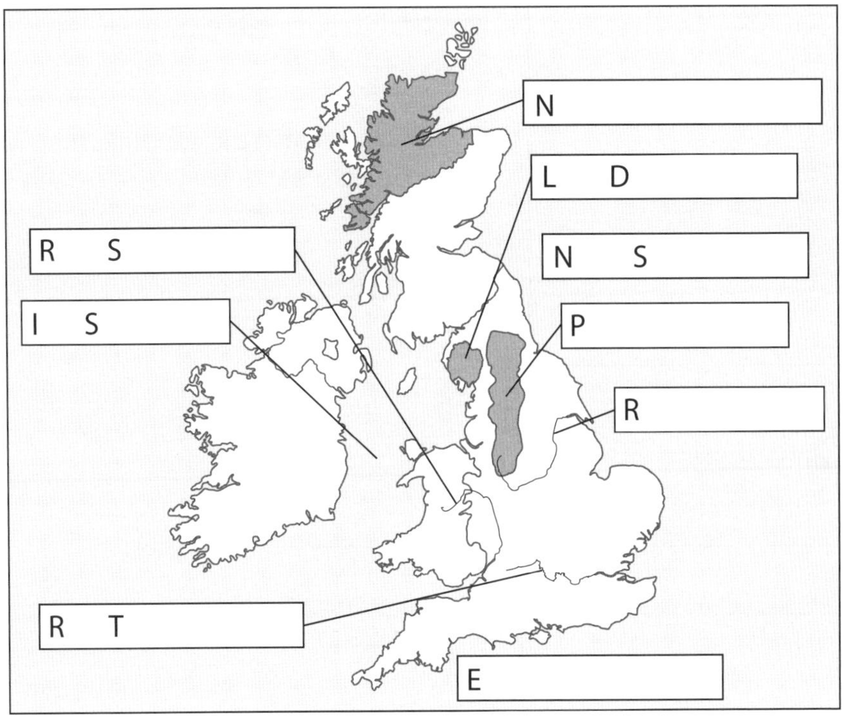

N
L D
R S
N S
I S
P
R
R T
E

3.2 It's a jigsaw!

This is about how we humans have carved up the British Isles.

geog.1

pages 48-49

1 Fill in the gaps using words from the box.

The British Isles is divided up into two countries: the United Kingdom and the

____________________.

The ____________________ in turn is made up of different nations: England,

____________________, Wales and Northern Ireland.

United Kingdom	Germany	British Isles
Republic of Ireland	Scotland	England

2 Now look at the map and answer these questions.

A
B
C
D
E

a A is called

____

b B is called

____

c D is called

____

d A–D together are called

____

e A–E together are called

____

f Finally, shade in the countries that make up Great Britain.

3 Draw a line on the map to the country or region that you live in and label it.
What is special about your country or region?

____

____

3.3 What's our weather like?

This is about the difference between weather and climate – and how the climate varies across the UK.

4th edition
geog.1

pages 50-51

1 Fill in the gaps.

- __________ means the state of the __________. Is it warm? wet? windy?

  It changes from day to day.

- __________ is the __________ weather in a place.

weather	average	atmosphere	climate

2 a Cross out the wrong word in these sentences.
In general:

- It is colder/warmer in the north, because it is further from the equator.
- It is also colder/warmer on high land. Up a mountain the temperature falls/rises.
- But in winter, a cold/warm ocean current called the North Atlantic Drift cools/warms the west coast. So the east coast is the coldest/warmest part in winter.

b The maps below show average temperatures in summer and winter. Colour in the first map in shades of orange. Make the warmest areas darkest, and the coldest areas lightest.

c Now colour the second map in blue. This time, make the coldest areas darkest. And don't forget the key!

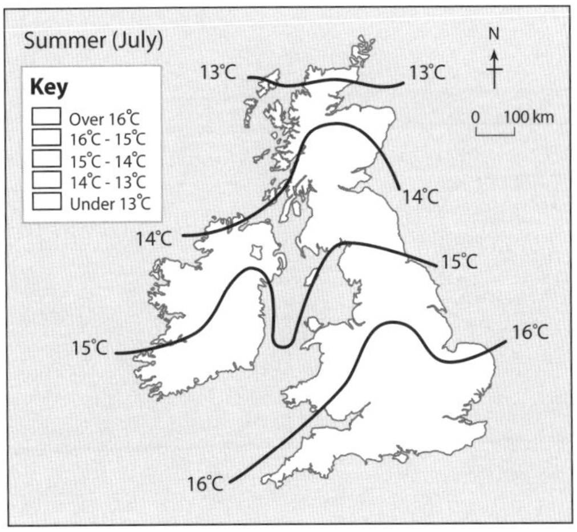
Summer (July)
Key
Over 16°C
16°C - 15°C
15°C - 14°C
14°C - 13°C
Under 13°C
N
0 100 km
13°C
13°C
14°C
14°C
15°C
15°C
16°C
16°C

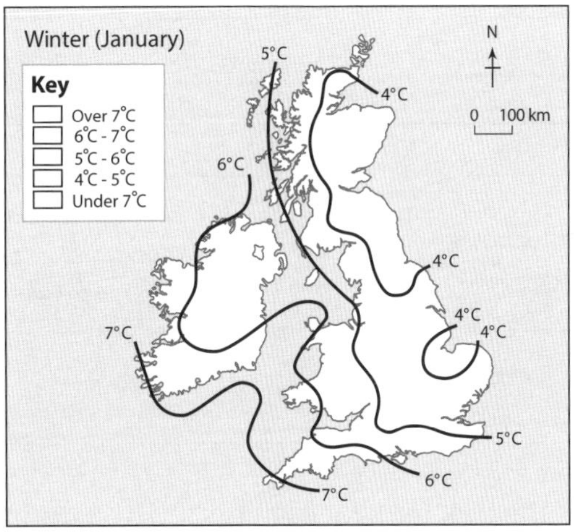
Winter (January)
Key
Over 7°C
6°C - 7°C
5°C - 6°C
4°C - 5°C
Under 7°C
N
0 100 km
5°C
4°C
6°C
4°C
4°C
4°C
7°C
5°C
6°C
7°C

3 Answer these questions in full sentences.

a Which parts of the British Isles are wettest?

____

b Can you explain why?

____

____

3.4 Who are we?

This is about how Britain has been peopled by immigrants.

4th edition
geog.1

pages 52-53

1 Some of the definitions below are incorrect, but which? Cross out any wrong terms and write the correct word in the second column. One has been done for you.

	Correct term
A An **asylum seeker** is a person who flees to another country for safety, and asks to be allowed to stay there.	
B An **invader** is someone who enters a country to attack it.	
C An ~~**emigrant**~~ is a person who comes into a country to live.	immigrant
D A **settler** is a person who takes over land to live on, where no one has lived before.	
E A **refugee** is a person who has been forced to flee from danger.	
F An **asylum seeker** is a person who leaves his or her own country to settle in another country.	
G A **refugee** is a person who moves to another part of the country or another country, often just to work for a while.	

2 Here are some statements from people who've arrived in the British Isles in the last 2000 years.

It's 48 AD. I am a centurion with the Roman army. Our aim is to expand our empire.

____

I came to live here a few years ago, in 2001. I didn't want to leave Kosovo but the war meant it was too dangerous to stay.

____

It's 1956. I've come here from Jamaica in search of a job.

____

It's 4000 BC. I've come here from Europe with my tribe. We're looking for a good place to farm.

____

Choose what you think is the best term for each person (use terms from question 1).

3 Do you agree with this statement: 'In the British Isles, we are all immigrants'?

____

____

____

3.5 Where do we live?

This is about how we humans have shaped the country, through where we chose to live!

4th edition
geog.1
pages 54-55

1 Fill in the gaps.

The __________ __________ of a place is the average number of people per square kilometre.

2 Tick the correct answer.

a The nation with the highest population density is …

England ☐ Wales ☐ Scotland ☐

b Of these areas, the one with the lowest population density is …

Cumbria ☐ Greater Manchester ☐ Devon ☐

c Of these cities, the largest is …

Edinburgh ☐ Birmingham ☐ Glasgow ☐

3 Look at this pie chart for the United Kingdom.

Where the UK population lives

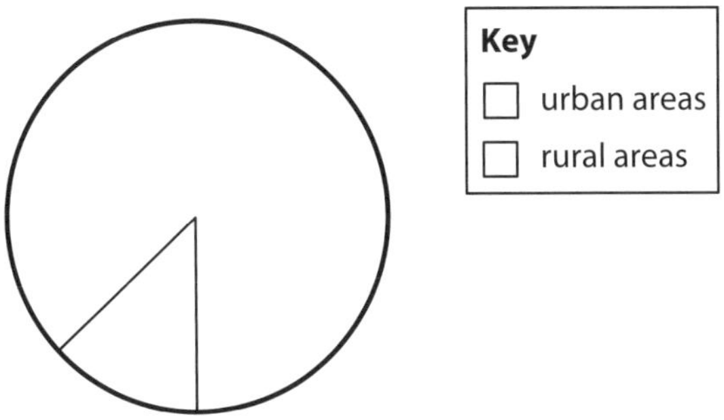
Key
urban areas
rural areas

a Shade in the chart and the key to show where the population of the UK lives.

b Imagine you live on a farm half an hour's drive to the nearest town.
Give three good points and three bad points about living in such a rural area.

Good points

____

____

____

Bad points

____

____

____

3.6 How are we doing?

This is about how some parts of Great Britain are better off than others – and some of the reasons why.

geog.1

pages 56-57

House prices, wages and unemployment vary from region to region in the UK.

Region	average house price (£)	average wage (£ per week)	unemployment (%)
A Wales	164 200	453	8.0
B Scotland	185 000	498	7.3
C Northern Ireland	129 700	460	7.3
D North East	149 400	455	10.3
E North West	165 600	470	8.6
F Yorks & Humber	167 300	465	8.8
G East Midlands	177 650	465	7.7
H West Midlands	187 800	470	9.4
I East	257 000	495	5.9
J London	436 550	650	8.6
K South East	304 000	540	6.0
L South West	230 400	467	6.3

* 2013 figures

1 Complete the bar chart below for house prices and wages in 2013. Wales has been done for you.

Average house price
Average wage
Average house price (£1,000)
Average wage per week (£)
0
100
200
300
400
500
600
700
A B C D E F G H I J K L

2 What do the differences in house prices tell you about inequality in the UK?

____

____

3 What does the relationship between house prices and wages tell you about inequality in the UK?

____

____

4 With the unemployment figures in mind as well, which region of the UK do you think would be the best to live in for income, employment and affordable housing? Why?

____

____

3.7 London, your capital city

This is about the London marathon. Can you work out its route from a description?

4th edition
geog.1

pages 58-59

The London marathon is the most popular marathon in the world. In 2013 there were 35 000 runners.

1 Read this description of the route of a recent London Marathon. Then trace the route of the race on the map below. Write in the words 'Start' and 'Finish'. Write in the missing words in the description and insert the numbers in the correct places on the map.

2046

The race starts in Greenwich Park, south of the river and just to the east of the well-known loop of the Thames at the Isle of Dogs. The runners head east, across the A2 to ______________ park. (1) At the A ______________ (2) road the race turns north, towards the river and then turns west along ______________ Church Street (3). They pass Maryon ______________ (4) and run along the A206, over the A2 again and towards Greenwich once more where they run past the famous ship, the Cutty ______________ (5). They then turn north-west, until they get to the old docks at Surrey ______________ (6). They then follow the loop of the river past the ______________ Tunnel (7). Heading west along the river past Bermondsey, they cross the Thames at ______________ Bridge (8). They now head east again past Wapping, and turn south into the Isle of Dogs, running along the Westferry Road. At the southernmost point of the loop they turn north, running towards the well-known skyscraper of Canary ______________ (9) Near the ______________ Tunnel (10) they turn west, heading for Limehouse. They then follow the Thames. They are now in the old City of ______________ (11). Still heading west they run past Waterloo Bridge. From here they can see the National Theatre and the Royal Festival Hall on the ________ ________ (12). At ______________ near the Houses of Parliament, they head away from the river towards St James Park. They run clockwise around the lake and the race ends at the northern end of the park.

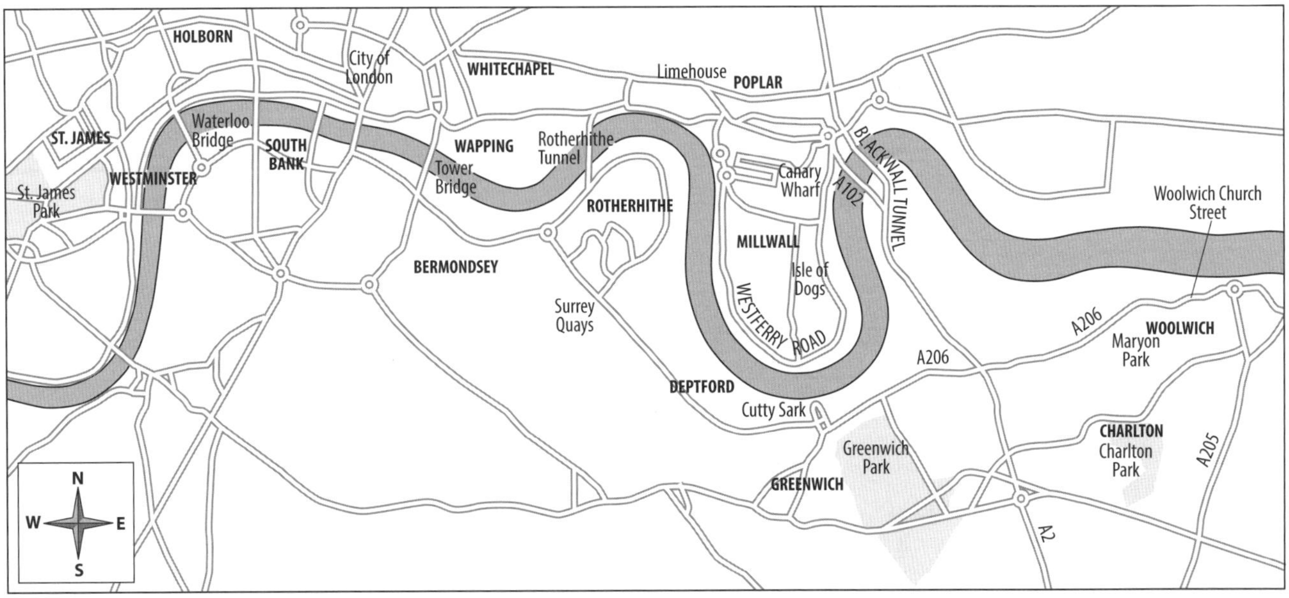

HOLBORN
City of London
WHITECHAPEL
Limehouse
POPLAR
ST. JAMES
Waterloo Bridge
SOUTH BANK
WAPPING
Rotherhithe Tunnel
Tower Bridge
Canary Wharf
BLACKWALL TUNNEL
A102
St. James Park
WESTMINSTER
ROTHERHITHE
Woolwich Church Street
MILLWALL
Isle of Dogs
BERMONDSEY
WESTFERRY ROAD
Surrey Quays
A206
WOOLWICH
Maryon Park
A206
DEPTFORD
Cutty Sark
CHARLTON
Charlton Park
Greenwich Park
A205
GREENWICH
A2
N
W
E
S

4 Glaciers

4th edition
geog.1
pages 60-61

Find the best photo of a glacier you can, and stick it in this space!

1 Write down two things you know about glaciers.

a ______________________________

b ______________________________

2 Write down two things you'd like to find out about glaciers.

a ______________________________

b ______________________________

At the end of this topic, come back and see if you've found out about these things. If you have, draw a ☺ next to your question – if you haven't, draw a ☹!

4.1 Your place ... 20 000 years ago!

This is about understanding when the British Isles was in the grip of ice.

geog.1

pages 62-63

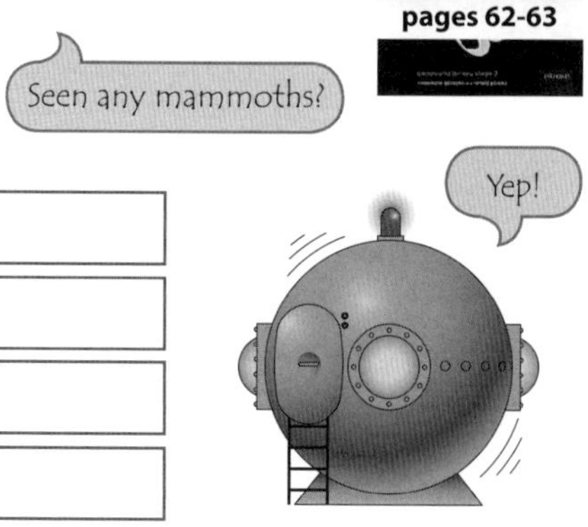

Seen any mammoths?
Yep!

1 Write 'True' or 'False' in the box after each of these sentences.

- **a** Woolly mammoth once roamed southern Britain.
- **b** Woolly mammoth were like very large sheep.
- **c** A few woolly mammoth can still be found in remote parts of Scotland.
- **d** Woolly mammoth were like hairy elephants.

2 Finish this timeline. The good thing is that the 'time' labels are already in place. To finish it, you need to add notes at each time label, saying what was happening at that time.

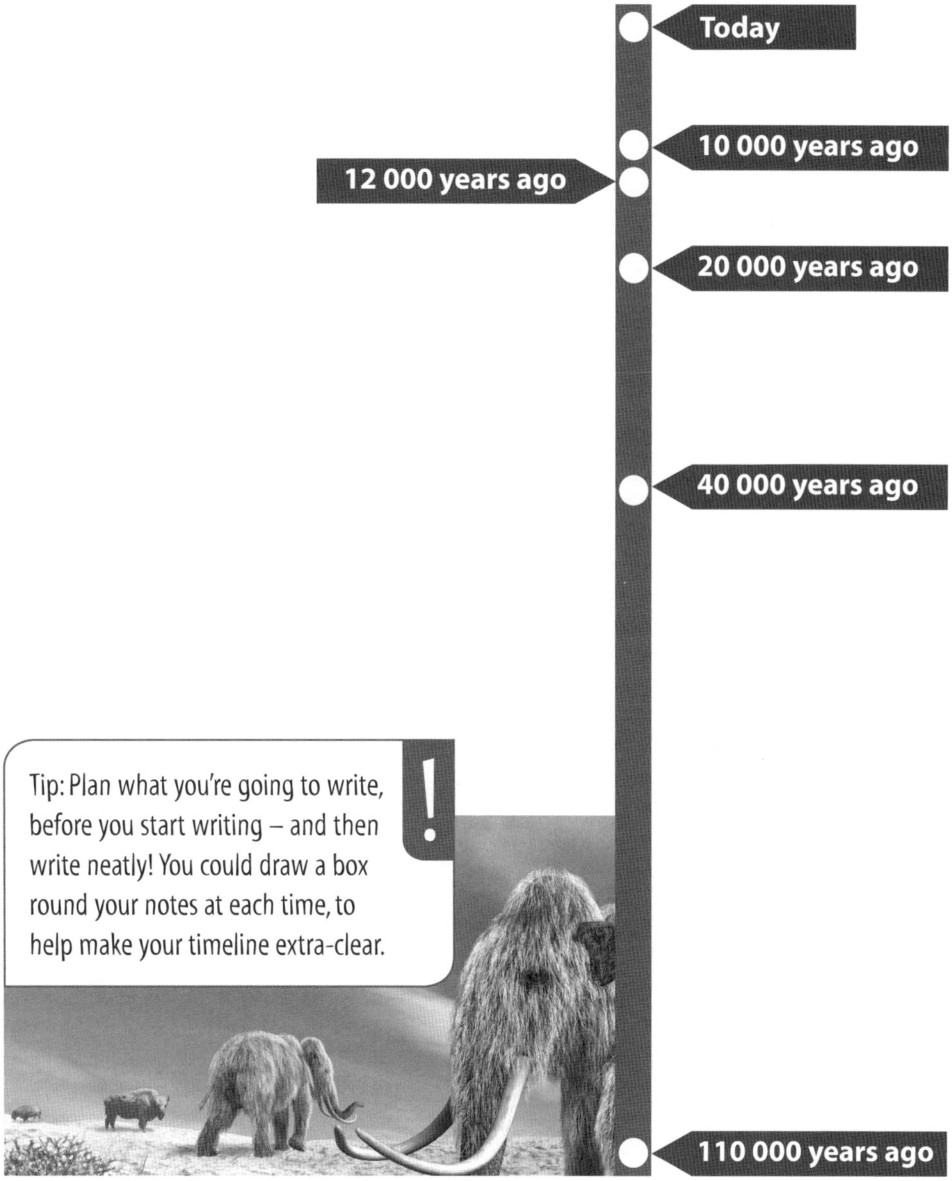

Today
10 000 years ago
12 000 years ago
20 000 years ago
40 000 years ago
110 000 years ago

Tip: Plan what you're going to write, before you start writing – and then write neatly! You could draw a box round your notes at each time, to help make your timeline extra-clear.

4.2 Glaciers

4th edition
geog.1
pages 64-65

This is about the world's glaciers today.

1 Tick the correct answer:

a How much of the Earth's surface do glaciers cover?

about 40% ☐ about 30% ☐ about 20% ☐ about 10% ☐

b During the last ice age, how much of the Earth's surface was covered by glaciers?

about 43% ☐ about 33% ☐ about 25% ☐ about 20% ☐

c Today, how much of the world's ice is in Antarctica and Greenland?

less than 90% ☐ 95% ☐ 99% ☐ over 99% ☐

d What are large cracks in glaciers are called?

cravats ☐ crevasses ☐ crevices ☐ creases ☐

e How many continents have glaciers?

two ☐ three ☐ five ☐ all seven ☐

2 Do some research to find out about Vatnajokull Glacier in Iceland.

Find a photo of Vatnajokull Glacier and stick it in the big box below, and then write a fact about Vatnajokull in each of the smaller boxes.

4.3 Glaciers at work

This is your chance to show that you know how glaciers shape the landscape.

geog.1
pages 66-67

1 Draw a spider diagram to show the work that glaciers do and how they do it. The first one has been started for you.

> Tip: A good way to do this would be to start with the three jobs glaciers do.

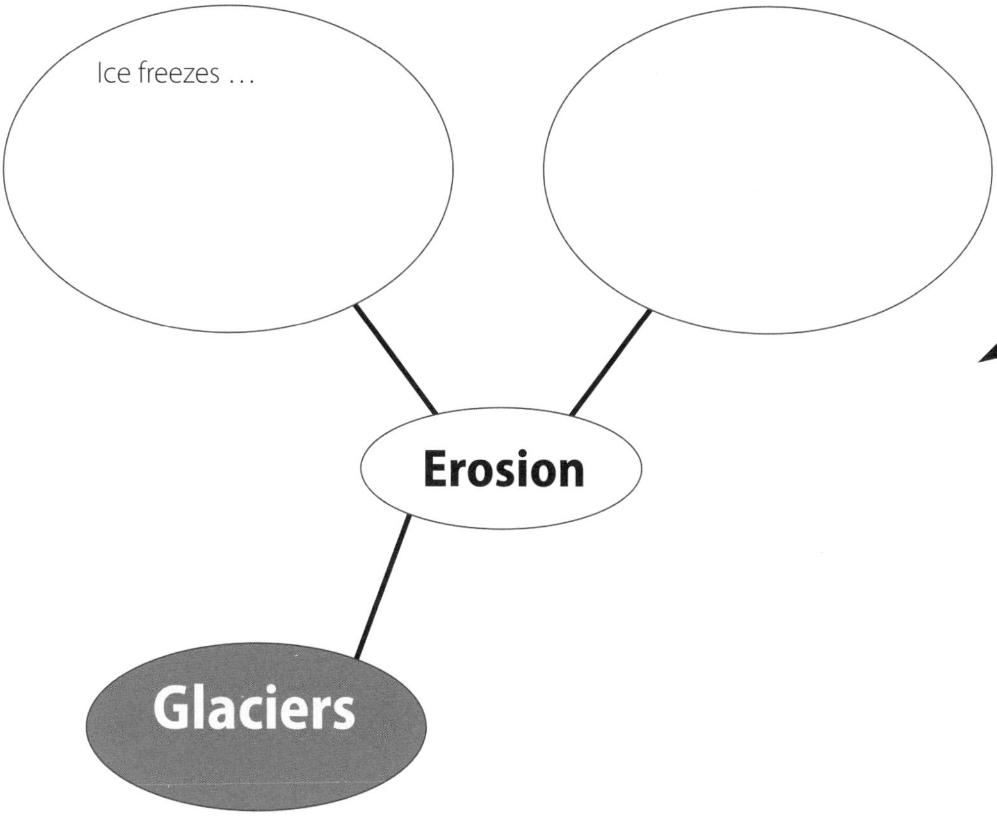

Ice freezes …
Erosion
Glaciers

> Tip: You could colour-code the different parts of your spider diagram – this would help make it even clearer.

4.4 Glacial landforms created by erosion: part 1

This is your chance to learn how glaciers can change the landscape.

geog.1
pages 68-69

1 The paragraph below describes how a corrie is formed.
Circle the correct word from each pair.

As snow falls, it **compacts / constructs** into ice. Through **abrasion / corrosion** the hollow gradually becomes **bigger / smaller** and the walls steeper. Eventually, with the help of freeze-thaw, the glacier is big enough to flow over the **edge / bottom** of the corrie and move down the mountain. When the glacier melts, the corrie often has a **lake / river** in it. This is often called a tarn.

2 Imagine you are going on an expedition to explore Bleaberry Tarn in the Lake District, shown in Photo A on page 68 in the pupil book. Write down five things that you think you would need to take to help make your expedition a success.

Give a reason for each of your answers.

I would need to take...	because...
1	
2	
3	
4	
5	

3 Study the photograph of Bleaberry Tarn again. Imagine you work for the Lake District Tourism Board. You have been asked to write about fifty words to describe what the photograph shows.

Tip: Remember to use as many adjectives (describing words) as you can.

4.5 Glacial landforms created by erosion: part 2

More ways in which glaciers can change the landscape!

geog.1

pages 70-71

1 Look at photos A and B. Write down three differences between the valleys shown in each photo.

Tip: Think about the shape of the valleys. **!**

A

B

1 ______________________________

2 ______________________________

3 ______________________________

2 Look at the map of the Lake District shown below. In your own words describe three facts about the lakes and where they are found. Remember to use the north arrow to help you.

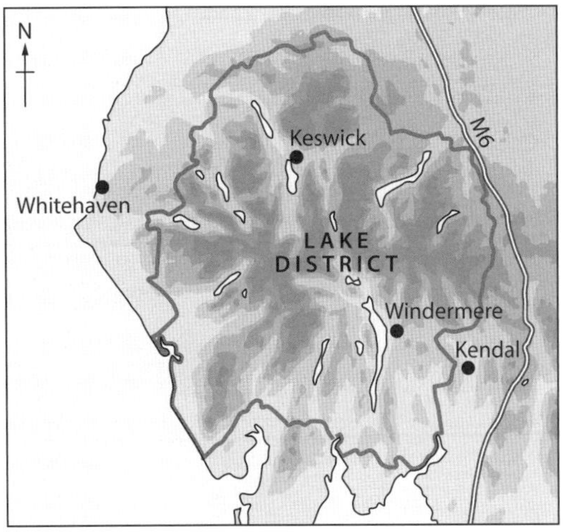

N
Keswick
Whitehaven
M6
LAKE DISTRICT
Windermere
Kendal

1 ______________________________

____

____

____

2 ______________________________

____

____

____

3 ______________________________

____

____

4.6 Glacial landforms created by deposition

This is where you learn about landforms created when glaciers melt.

geog.1
pages 72-73

1 Glaciers form many landforms, even when they melt! Complete the fact sheet below that describes and explains the landforms that are created.

When glaciers melt they change the landscape

A terminal moraine is...

It is caused when...

An erratic is...

They are found in strange places because...

Drumlins are...

They are smooth-shaped because...

A lateral moraine is...

It is caused when...

2 Write and present a two minute talk for your class about one of the landforms created by glaciers – either through glacial erosion or deposition. You should produce one diagram or PowerPoint slide to help you.

4.7 Glacial landforms on an OS map

This is your chance to become a detective to find glacial landforms on an OS map.

4th edition
geog.1

pages 74-75

1 Here is the description of a place shown on the map in the pupil book on page 75. Where is it? Use the clues to help you.

I am to the west of Crummock Water. Mosedale Beck flows in a valley to my east, the waterfall called Scale Force is found south-east of me. I am 509 metres high, and my name would make feathers fly!

What is my four figure grid reference?

The place is called ____________________

The four figure grid reference for it is ____________

2 Now write your own clues to describe a place or feature shown on the map. Then test them on a partner. Can they identify it using your clues?

3 Use the clues below to follow a route around the area shown by the map. Write the places that you reach in the blank spaces in the paragraph. Remember to use the clues and the scale to help you!

I start walking from the car park at the Gatesgarth at 6-figure grid reference __________. I follow the road north-west until I get to the hotel in the village of __________ in grid square 1716. Here, I take the footpath on my left that crosses the valley floor and goes through the forest called __________ Wood. I go past Bleaberry Tarn and climb upwards towards __________ Forest. Here, I turn left and follow the long footpath south-east alongside the stream until I get to a building with a pirate name, called __________ __________ __________ at 6-figure grid reference __________.

4.8 Glaciers and us

This is your chance to consider if glaciers really matter.

4th edition
geog.1

pages 76-77

1 Pages 76 and 77 in the pupil book tell us about five facts about glaciers. look at the list below and put them in rank order of importance, with 1 being the most important.

Facts about glaciers	Rank order
Glaciers bring in tourists.	
Glaciers present a challenge.	
Glaciers support life.	
Glaciers are in need of protection.	
Glaciers warn us about climate change.	

2 Write three sentences explaining why you have chosen your number 1. Remember to give reasons for choosing it!

Sentence 1 ______________________________

____

Sentence 2 ______________________________

____

Sentence 3 ______________________________

____

3 Design a poster in the space below showing what you know about glaciers. Use all the information that you have learned in this unit to help you.

5 Rivers

4th edition
geog.1
pages 78-79

1

A 1	**T** 2	**I** 1	**D** 1	**S** 3	**R** 2	**C** 3	**N** 2
L 2	**E** 1	**V** 5	**U** 3	**F** 3	**O** 1	**W** 4	**P** 6

Look at the letter squares above. Each letter has a value. Use the letters to make as many words that you can that are to do with rivers. You can use each letter more than once. Which of your words is worth the most? What is the longest word you can make? Challenge others in your class!

Tip: Use page 79 in the pupil book to help you. **!**

2 In the space below, write a paragraph about rivers. You must use all of the words that you found in your answer to the question above.

5.1 Meet the River Thames

This is where you will learn more about changes along England's longest river.

geog.1 pages 80-81

1 The River Thames starts its life at Thames Head in the Cotswolds. Research what the river is like as it flows through each of the other places shown on the map below. Complete the table with the information you find. Show how the river looks different in the three places.

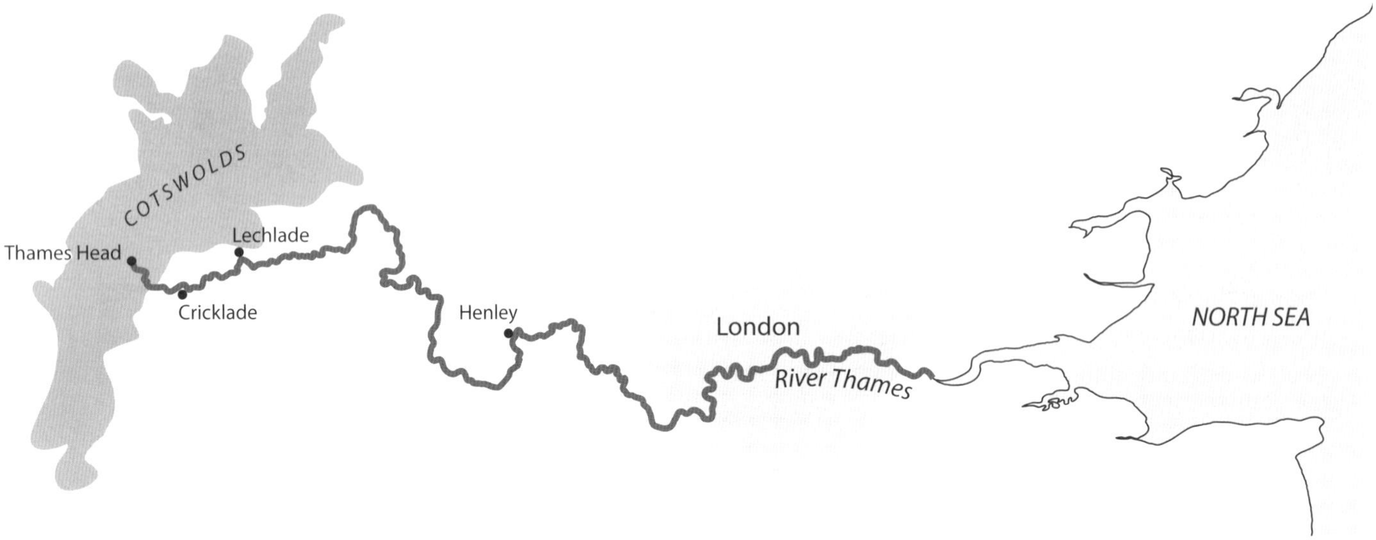
COTSWOLDS
Thames Head
Lechlade
Cricklade
Henley
London
River Thames
NORTH SEA

When the River Thames flows through …	… it is like this:
Cricklade	
Lechlade	
Henley	

5.2 It's the water cycle at work

This is about the water cycle, and how rainfall reaches a river.

4th edition
geog.1

pages 82-83

1 Water moves between the ocean, the air and the land. This circulation is called the water cycle.

a Fill in the gaps below, choosing words from the box (you don't have to use them all).

☐ The air ____________. High up, where it's cooler, the water vapour ____________ into tiny water droplets. These form ____________.

☐ The water drops fall as rain (or hail or sleet or snow).

☐ The sun warms oceans, lakes and seas, turning water into water vapour. This is called ____________.

☐ Some water runs along the ground, and some soaks through it, heading for streams and rivers.

☐ The droplets inside the clouds grow into larger droplets, leading to ____________.

☐ The river carries the water back to the ____________. The ____________ is complete. And then it starts all over again…

evaporation	rises	infiltration
gas	precipitation	clouds
condenses	ocean	cycle

b Add numbers in the small boxes so that the sentences are in the correct order.

2 Draw a diagram in the box below to show how rainwater reaches a river. Try to use as many of these words as possible (there are a few clues to help you!):

interception (when rainwater catches leaves)
surface runoff
throughflow
groundwater
groundwater flow
infiltration (the soaking of rainwater into the ground)
permeable (lets water soak through)
impermeable

5.3 A closer look at a river

This is about the different parts of a river.

geog.1
pages 84-85

1 **a** Cross out the incorrect word in these sentences.

- The point where two rivers join is called a tributary/confluence.
- The confluence/watershed is an imaginary line that separates one drainage basin from the next.
- The source/mouth is where the river flows into a lake, or the sea, or the ocean.
- The flat land around a river that gets flooded when the river overflows is the flood plain/tributary.
- The mouth/source is the starting point of the river.
- The land around a river from which water drains into the river is the river basin/watershed.

b Now draw a sketch map of an imaginary river. Try to mark on and label all the features from **1a**.

2 Fill in the gaps.

A drawing of the river's ____________________ ____________________

The __________ is the highest point.
The slope gets less steep in this middle stretch.
The __________ is the river's lowest point.
lake or sea
different layers of rock below the river
Now the slope is flattening out.

5.4 A river at work

4th edition
geog.1

pages 86-87

This is about how rivers shape the land, by picking up, carrying and dropping material.

1 Fill in the gaps choosing words from the box.

Rivers do their work in three stages:

1 They pick up or ____________ material from one place.

2 They carry or ____________ it to another place

3 Then they drop or ____________ it.

erode	transport	deposit

2 Finish off this cartoon to tell the story of Sid the Stone's journey. (You don't have to fill all the boxes.)

1

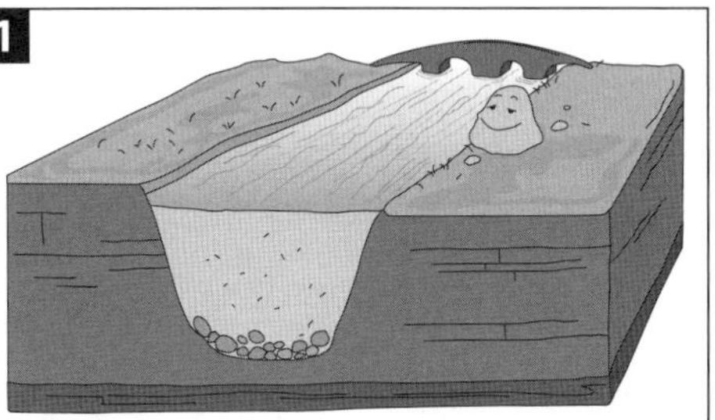

Sid the stone had lived in the river bank for as long as he could remember. Then, one day ...

2

… he was prised out of the bank by **hydraulic action!**

3

____

____

4

____

____

5

____

____

6

____

____

5.5 Five landforms created by the river

This is about the landforms a river creates, by eroding and depositing material.

geog.1

pages 88-89

1 Fill in the gaps in this table.

Landform	Description	Created by ...
V-shaped valley	a valley shaped like the letter V, carved out by a river	
waterfall		erosion
gorge	a narrow valley with steep sides	erosion
	a bend in a river	erosion + deposition
oxbow lake	a lake formed when a loop of river gets cut away	

2 **a** These pictures show how a waterfall develops.
Under each picture, describe what is going on.

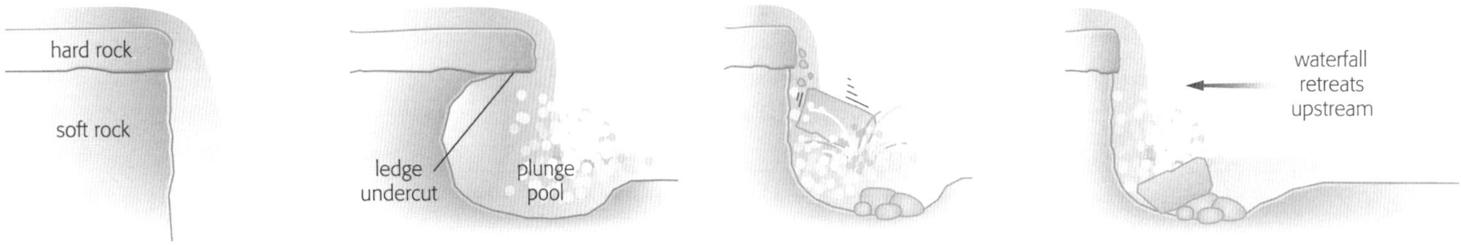

hard rock
soft rock
ledge undercut
plunge pool
waterfall retreats upstream

1 ______________________

2 ______________________

3 ______________________

4 ______________________

b Draw pictures in the boxes below to show how a meander develops.

1 Water flows faster on the outer curve of the bend, and slower on the inner curve. So …

2 … the outer bank gets eroded, but material is deposited at the inner bank. Over time …

3 … as the outer bank wears away, and the inner one grows, a meander forms.

4 As the process continues, the meander grows more 'loopy'.

5.6 Rivers and us

geog.1
pages 90-91

This is about how we make use of rivers and how sometimes we may damage them.

1 a Look at the statements below. Draw a line linking the statements that you think are connected. One has been done for you.

A Dams are built across some rivers	and then pumped to our taps
B Farmers pump water from rivers	and then sent back to rivers
C Water from rivers is cleaned	to irrigate their land
D Dirty water from our houses is cleaned	so that the water turns turbines to make electricity

b Choose one of your completed statements from A to D. Copy it out on the line below.

____

c Do you think what you have just written is good or bad for the river? Give reasons for your answer.

____

____

____

2 Read the speech bubbles below.

Rivers are for wildlife more than for people
Environmentalist

My fishing lines get caught by the boats and break
Fisherman

Without the water from the river my crops will not have enough water
Farmer

We are quiet and do not disturb anybody
Boat owner

There are very few places left where we can walk our dogs in peace
Dog walker

Choose the one person that you think may damage the river the most. Give reasons to explain your choice.

____

5.7 Rivers and our water supply

This is where you will find out how rivers help keep you alive!

4th edition
geog.1

pages 92-93

1 Cross out the incorrect word in these sentences.

- Across the UK water is pumped from rivers into lakes / reservoirs.
- A layer of rock that holds groundwater is called an aquaplane / aquifer.
- In a water treatment plant, chlorine is added to kill fish / germs.
- The clean water is then put in a storage reservoir and then from there it flows to rivers / homes.
- In a sewage plant the dirty water is cleaned up by bacteria / oxygen.
- Clean water is put into pipes / rivers.

2 Carry out a survey about how, and where, water is used in your school. Write down five of your findings in the space below.

Use the results to create an advert for your school website encouraging *either* adults *or* children to save water.

> Tip: Remember who your target audience is, and be as persuasive as you can. **!**

How is water used in school?	Where is it used?
1	
2	
3	
4	
5	

5.8 Floods!

This is where you will find out how and why rivers flood.

geog.1 pages 94-95

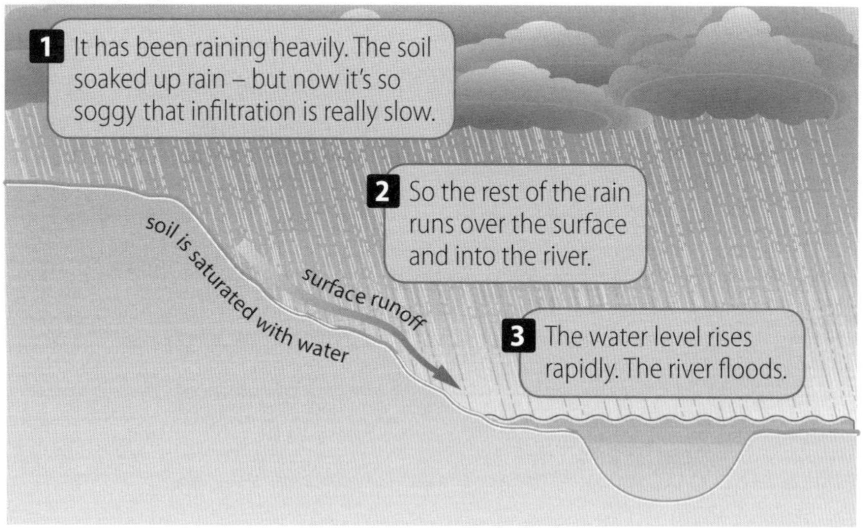
1 It has been raining heavily. The soil soaked up rain – but now it's so soggy that infiltration is really slow.
2 So the rest of the rain runs over the surface and into the river.
3 The water level rises rapidly. The river floods.
soil is saturated with water
surface runoff

1 **a** Look at the diagram above. In your own words explain what is meant by the following words and phrases.

Infiltration ______________________________

____

____

____

Surface runoff ______________________________

____

____

____

2 When floods happen people often talk about three things: flood prevention, flood defence and flood warning. In the spaces below write down what you think is meant by each term.

Flood prevention: ______________________________

Flood defence: ______________________________

Flood warning: ______________________________

3 Which of the three do you think is the most important? Give reasons why.

I think ______________ is the most important because ______________________________

____

____

5.9 Flooding on the River Thames

This is where you will think about the impacts of flooding on the River Thames.

geog.1

pages 96-97

1 Look at the photographs on page 96 in the pupil book that show flooding in Oxford, Abingdon and Twickenham. Imagine that was your house...and your local shop!

Describe, in the space below, how you and your family would cope with the floods.

> Tip: Think about how it would change your daily life. !

2 Write down five things that you think it would be essential to save if your house got flooded. Explain your answers.

I would save ...	because ...
1	
2	
3	
4	
5	

3 a Look at the statements below that show what people living by the River Thames may do if their area became flooded. Put them in priority order, by writing the numbers 1-5 by each. Number 1 would be the first thing that you think they should do, and 5 the last.

Ring family		Phone 999		Turn off power		Save pets		Move upstairs	

b In the space below give reasons to explain your first choice.

My first choice is

This is because

5.10 Protecting ourselves from floods

geog.1

pages 98-99

This is where you will think about the impacts of flooding on the River Thames.

1 Read this information about London. Do you think London should be protected by another Thames Barrier? Fill in the speech bubbles below to show your thoughts.

- *London's population is 8.17 million (2011 data) and growing*
- *Architect Sir Terry Farrell has proposed building a five-mile barrage from Southend to Sheerness - basically the mouth of the Thames - all linked by islands.*
- *A second barrier could house turbines and use tide flows to generate electricity for London.*
- *The Farrell proposal includes a road and rail bridge to create a transport link between the two counties and the islands could be used for housing and leisure facilities, the sale of which will help pay for it all.*
- *Another idea is to set aside large areas of open country downstream of London as emergency flood plains, so protecting the London.*
- *London's exports add up to about 24 per cent of the total value of all UK exports*
- *The Thames Barrier took eight years to build, costing £535m (£1600m at 2013 prices) and became fully working in 1982.*
- *Europe's largest shopping centre is in London - Westfield in Stratford, near the Olympic Park*
- *"The Thames Barrier was built in response to the floods in 1953. Nobody had heard of global warming then." Dr Richard Bloore (January 2013)*
- *In 2007 a second Thames barrier was costed at £20 billion.*
- *The Thames Barrier was originally designed to work until the year 2030.*
- *There are 270 stations on London's underground tube network*
- *3.4 million visitors came to London in the first three months of 2013*
- *The average house price in London (May 2013) is £437,000*
- *Climate change is causing sea level rise, which could mean larger storm surges in the Thames.*
- *The Thames Barrier currently protects 125sq km of London, including an estimated 1.25m people, £80bn worth of property, a large proportion of the London tube network and many historic buildings, power supplies, hospitals and schools.*

What is the issue?

What should happen?

What is the risk?

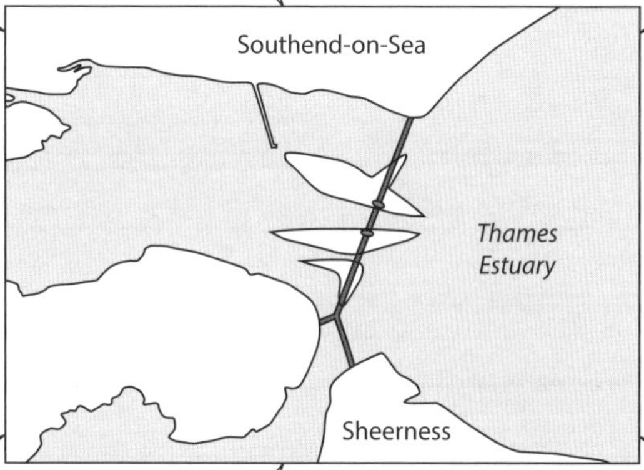
Southend-on-Sea
Thames Estuary
Sheerness

Who/what will lose?

What is the options?

Who/what will benefit?

6 Africa

geog.1
pages 100-101

1 Test yourself. Without looking anywhere else, draw an outline of Africa in the box above. Then draw in the borders of all the countries you can think of. Include the names of cities, mountains, rivers, lakes and deserts.

2 Write down all you know about three of the places you have put on the map.

Place 1 ______________________________

____

Place 2 ______________________________

____

Place 3 ______________________________

____

3 At the end of this topic, come back and see if you've been correct about these places. Draw a ☺ next to each place you got more or less right. If you were wrong draw a ☹ next to the place!

What and where is Africa?

This is about locating the continent of Africa on the world map.

geog.1

pages 102-103

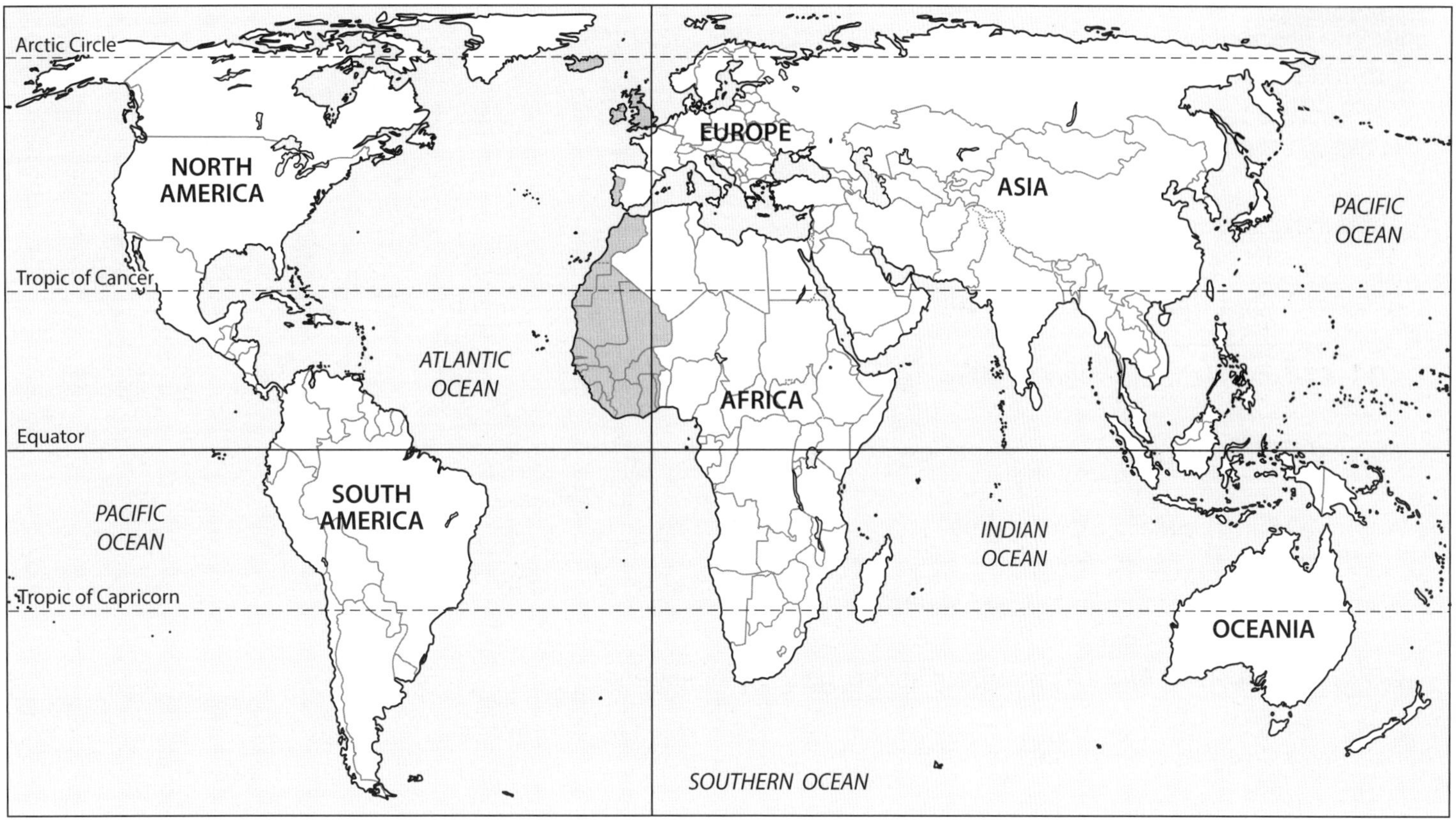

0°
Arctic Circle
Tropic of Cancer
Equator
Tropic of Capricorn
NORTH AMERICA
SOUTH AMERICA
EUROPE
ASIA
AFRICA
OCEANIA
PACIFIC OCEAN
ATLANTIC OCEAN
PACIFIC OCEAN
INDIAN OCEAN
SOUTHERN OCEAN

1 Colour in all the land in the world that lies between the Tropic of Cancer and the Tropic of Capricorn. This area is known as the 'tropics' and is the hottest part of the world.

a Which continent has most of its land area inside the tropics? ____________________

b Which continent lies completely outside the tropics? ____________________

2 Using a second colour fill in all the land area of Africa that lies outside the tropics. What proportion of Africa lies outside the tropics? Circle the correct answer.

A About 60% B About 75% C About 35%

3 The map also shows 0° longitude, which is known as the Prime Meridian. It runs through Greenwich in London. The countries that share the Greenwich Meridian Time Zone (GMT) with the UK have been shaded on the map.

a How many of the countries in Africa which share the same time zone as us can you name?

____

____

6.2 A little history

This is an exercise in placing key events in African history into the correct order.

4th edition
geog.1

pages 104-105

1 The list of ten key events in African history below has been jumbled up. Put the list into the right order in which the events occurred by writing a number from 1 to 10 in each box.

About AD 1400.The Kingdom of Kongo begins. ☐

About 60 000 years ago. Homo Sapiens leaves Africa. ☐

About AD 1800. The Atlantic Slave Trade is ended. ☐

AD 1951. Libya is the first African colony to gain independence. ☐

About 2 million years ago. The first species of human appears. ☐

About AD 800. The Mali Empire begins. ☐

About 200 000 years ago. Homo Sapiens appears. ☐

AD 1980. Zimbabwe, Britain's last African colony, gains independence ☐

AD 1420. Portuguese exploration of Africa begins. ☐

About 3000 BC. The Ancient Egyptian civilization begins. ☐

2 Two African countries, shaded on the map below, were never European colonies. Name these two countries and write about them below.

Country A ____

____

____

____

Country B ____

____

____

____

A
B

6.3 Africa today

geog.1

pages 106-107

This is about looking at population growth figures and asking questions about them.

The population of Africa is expected to double in the next 35 years.
But the figures for the whole of Africa do not tell the full story.

1 The table below shows the population figures (in millions) for six of the largest African countries for 1950 and 2013 and the projected figures up to 2100.

	1950	2013	2025	2050	2100
Democratic Republic of Congo	12	68	92	155	262
Egypt	22	82	97	122	135
Ethiopia	18	94	125	188	243
Nigeria	38	174	240	440	913
South Africa	14	53	57	63	64
Tanzania	8	49	69	129	275

* UN figures

a Which country grew, or is projected to grow the fastest between;

1950- 2013 ____________________? 2013-2100 ____________________?

b Which country grew or is projected to grow the slowest between;

1950-2013 ____________________? 2013-2100 ____________________?

c Which country is projected to grow faster between 2013-2100 than it has grown between 1950-2013? ____________________.

2 a These figures are a 'medium variant' estimate; they could go up or down. What factors could cause these figures to vary? ____________________

____

b The projected figures for South Africa show a very small increase in population. Can you give an explanation for this? ____________________

____

c What needs to happen for the projected figures of the other countries show a similar decline in growth? ____________________

____

3 Use the map below to work out the official languages of the following countries:

A Angola ____________ B Chad ____________

C Equatorial Guinea ____________ D Gabon ____________

E Ghana ____________ F Uganda ____________

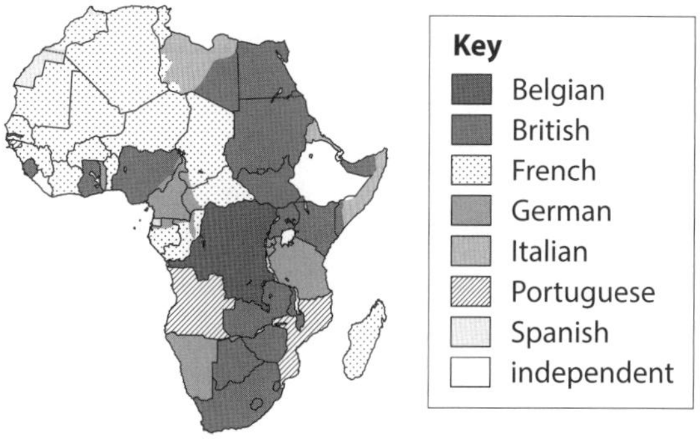
Key
Belgian
British
French
German
Italian
Portuguese
Spanish
independent

6.4 Africa's countries

This will help you understand Africa's size and the variety of its countries.

4th edition
geog.1

pages 108-109

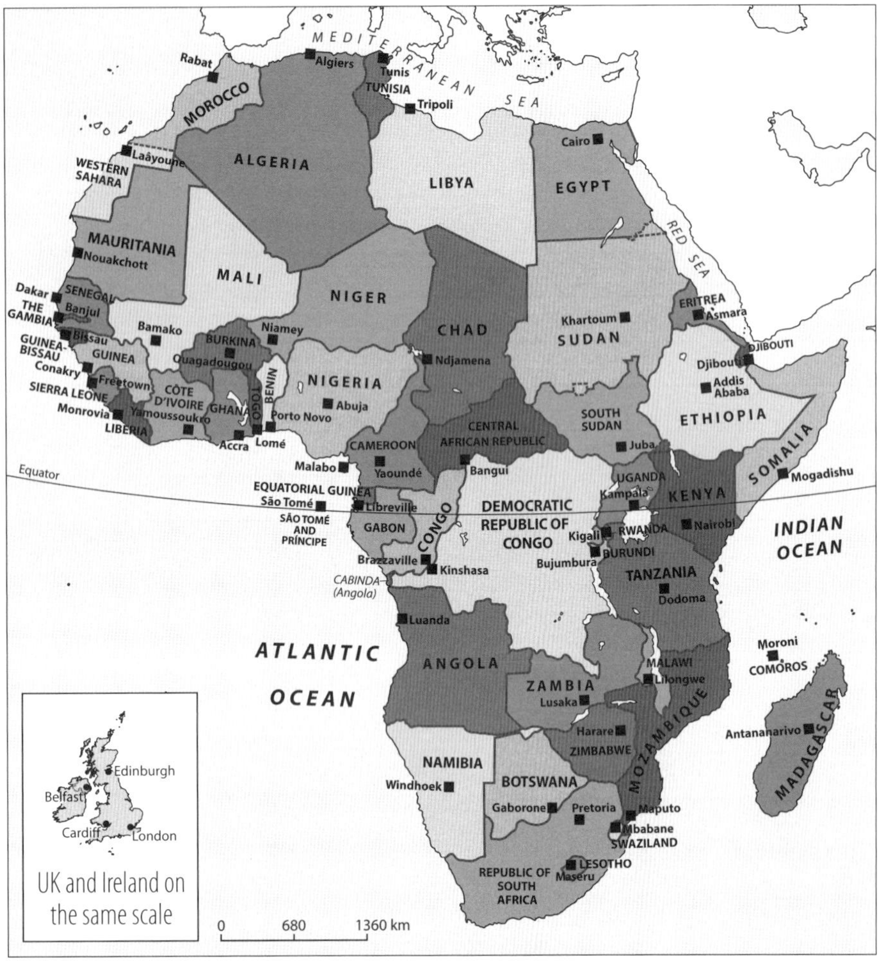

MEDITERRANEAN SEA
Rabat
Algiers
Tunis
TUNISIA
Tripoli
MOROCCO
Laâyoune
WESTERN SAHARA
ALGERIA
LIBYA
Cairo
EGYPT
RED SEA
MAURITANIA
Nouakchott
MALI
NIGER
CHAD
Khartoum
SUDAN
ERITREA
Asmara
Dakar
SENEGAL
Banjul
THE GAMBIA
Bissau
GUINEA-BISSAU
GUINEA
Conakry
Bamako
Niamey
BURKINA
Ouagadougou
BENIN
TOGO
NIGERIA
Abuja
Ndjamena
DJIBOUTI
Djibouti
Addis Ababa
ETHIOPIA
Freetown
SIERRA LEONE
Monrovia
LIBERIA
CÔTE D'IVOIRE
Yamoussoukro
GHANA
Accra
Lomé
Porto Novo
CENTRAL AFRICAN REPUBLIC
SOUTH SUDAN
Juba
SOMALIA
Mogadishu
Equator
Malabo
CAMEROON
Yaoundé
Bangui
EQUATORIAL GUINEA
São Tomé
SÃO TOMÉ AND PRÍNCIPE
Libreville
GABON
CONGO
DEMOCRATIC REPUBLIC OF CONGO
UGANDA
Kampala
KENYA
Kigali
RWANDA
Nairobi
BURUNDI
Bujumbura
Brazzaville
Kinshasa
CABINDA (Angola)
TANZANIA
Dodoma
INDIAN OCEAN
Luanda
ATLANTIC OCEAN
ANGOLA
ZAMBIA
Lusaka
MALAWI
Lilongwe
Moroni
COMOROS
Harare
ZIMBABWE
MOZAMBIQUE
Antananarivo
MADAGASCAR
NAMIBIA
Windhoek
BOTSWANA
Gaborone
Pretoria
Maputo
Mbabane
SWAZILAND
LESOTHO
Maseru
REPUBLIC OF SOUTH AFRICA
Edinburgh
Belfast
Cardiff
London
UK and Ireland on the same scale
0
680
1360 km

Country	
Algeria	
Angola	
Benin	
Botswana	
Burkina Faso	
Burundi	
Cameroon	
Cape Verde	
Central African Republic	
Chad	
Comoros	
Côte d'Ivoire	
Dem. Rep. Congo	
Djibouti	
Egypt	
Equatorial Guinea	
Eritrea	
Ethiopia	
Gabon	
Gambia	
Ghana	
Guinea	
Guinea-Bissau	
Kenya	
Lesotho	
Liberia	
Libya	
Madagascar	
Malawi	
Mali	
Mauritania	
Mauritius	
Morocco	
Mozambique	
Namibia	
Niger	
Nigeria	
Republic of Congo	
Rwanda	
São Tomé and Príncipe	
Senegal	
Seychelles	
Sierra Leone	
Somalia	
South Africa	
South Sudan	
Sudan	
Swaziland	
Tanzania	
Togo	
Tunisia	
Uganda	
Zambia	
Zimbabwe	

1 Using the scale on the map above find out how far it is between the following African capital cities.

a Algiers and Pretoria ____________ **b** Cairo and Abuja ____________

c Addis Ababa and Bamako ________ **d** Mogadishu and Banjul ________

2 Compare distances in the British Isles with distances in Africa. Name two African capital cities that are about the same distance apart as:

a London and Cardiff ____________ **b** Belfast and Cardiff ____________

c London and Edinburgh ____________

d Into which African country would the island of Britain fit easily both north-south and east-west? ____________

3 Can you guess the size of Africa's countries? Just looking at the map and using a ruler try to guess what the order of size is of Africa's 54 countries. Write in numbers from 1 (largest) to 54 (smallest) in the boxes alongside the names of the countries. Compare your guesses with a partner's, then look up the answers to see who was the closest!

6.5 Population distribution in Africa

4th edition
geog.1
pages 110-111

This looks at population differences between African regions and selected countries.

More and more Africans are moving from the countryside to towns or big cities. Here are the rural and urban population percentages for the different African regions.

Region	Urban	Rural
Eastern	23.6%	76.4%
Central	43.1%	56.9%
Northern	51.2%	48.8%
Southern	58.7%	41.3%
Western	44.9%	55.1%

1 Complete the pie graphs below to show the rural and urban divisions within the African regions. One has been done for you.

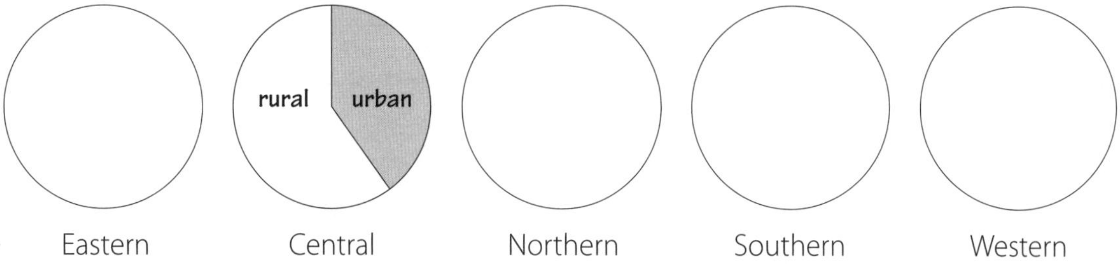
rural
urban
Eastern
Central
Northern
Southern
Western

Look back at the table in 6.3 showing population projections for six African countries. They represent all of the African regions.

2 Write down which regions they are in: Tanzania and Ethiopia ______________ Egypt ______________

Nigeria ______________ South Africa ______________ Democratic Republic of Congo ______________

3 The six countries in the table have the largest populations in their regions. So their urban/rural differences reflect the regions they are in. What links might there be between rural/urban differences and their rate of population increase?

Here are the percentages of the populations of the six countries that are under 14 (your generation!)

DRC = 45% Egypt = 31.1% Ethiopia = 42.7% Nigeria = 44.4% South Africa = 29.5% Tanzania = 44.9%

4 How might the proportion of the population that is under the age of 14 be connected to their rate of population increase?

5 Can you name any other factors that might be used in predicting the future population of a country?

6.6 Africa: physical features

Here we look at the relationship between Africa's physical features and its countries.

geog.1
pages 112-113

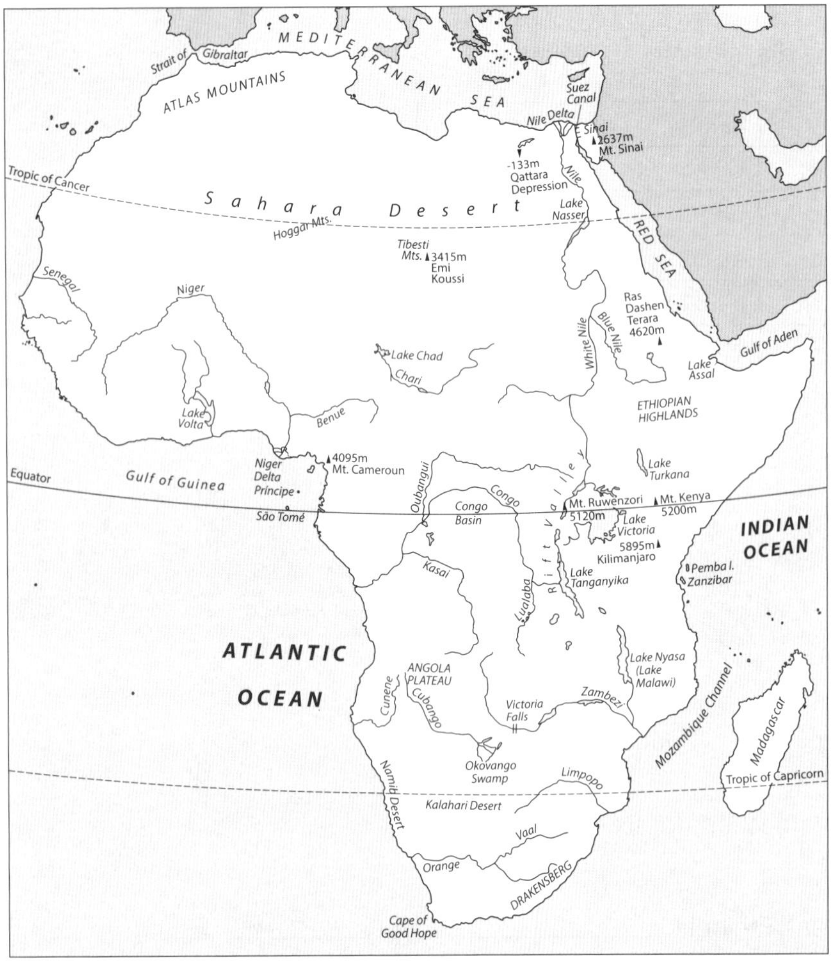

MEDITERRANEAN SEA
Strait of Gibraltar
ATLAS MOUNTAINS
Suez Canal
Nile Delta
Sinai
2637m Mt. Sinai
-133m Qattara Depression
Nile
Tropic of Cancer
Sahara Desert
Hoggar Mts.
Lake Nasser
RED SEA
Tibesti Mts. 3415m Emi Koussi
Senegal
Niger
Ras Dashen Terara 4620m
White Nile
Blue Nile
Lake Chad
Chari
Lake Assal
Gulf of Aden
ETHIOPIAN HIGHLANDS
Lake Volta
Benue
Niger Delta
4095m Mt. Cameroun
Equator
Gulf of Guinea
Principe
São Tomé
Oubangui
Lake Turkana
Rift Valley
Congo Basin
Congo
Mt. Ruwenzori 5120m
Mt. Kenya 5200m
Lake Victoria
5895m Kilimanjaro
INDIAN OCEAN
Kasai
Lake Tanganyika
Pemba I.
Zanzibar
Lualaba
ATLANTIC OCEAN
ANGOLA PLATEAU
Cunene
Cubango
Lake Nyasa (Lake Malawi)
Zambezi
Victoria Falls
Mozambique Channel
Madagascar
Okovango Swamp
Limpopo
Tropic of Capricorn
Namib Desert
Kalahari Desert
Vaal
Orange
DRAKENSBERG
Cape of Good Hope

1 Compare this map with the political map on page 56 and answer the following questions;

a Which countries are the Atlas Mountains in?

____

b In which countries is the Kalahari Desert?

____

c In which country is Mount Kilimanjaro?

____

d In which country are the Tibesti Mountains?

____

e In which country does the River Benue rise?

____

2 On the map circle those places where rivers form the borders or part of the borders between African countries. Name these rivers and the countries either side of them. [Note that rivers sometimes form a part but not all of the border between countries].

3 What is unusual about the River Cubango?

____

____

____

4 Research and find out what is unusual about Lake Volta.

____

____

____

Africa's biomes

geog.1
pages 114-115

This about telling what biome a place is in from its weather statistics.

The letters on the map represent five cities;

Dodoma Tamanrasset Yaoundé Gaborone Addis Ababa

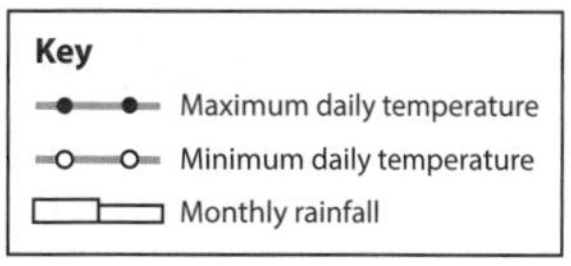
Key
Maximum daily temperature
Minimum daily temperature
Monthly rainfall

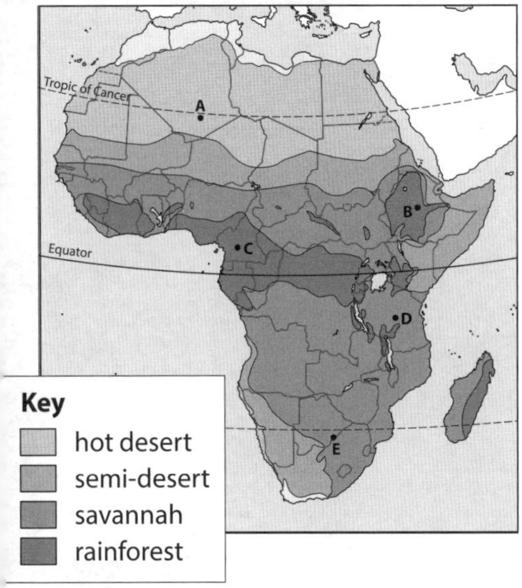
Tropic of Cancer
A
B
C
Equator
D
E
Key
hot desert
semi-desert
savannah
rainforest

Here are the climate charts for these places. Match each place on the map to a numbered chart and write down which biome it represents (hot desert, semi-desert, savanna, rainforest and mountain) Bear in mind the height above sea level (elevation) of each place when looking at the temperature figures. Also write down the reasons you matched that place to its biome.

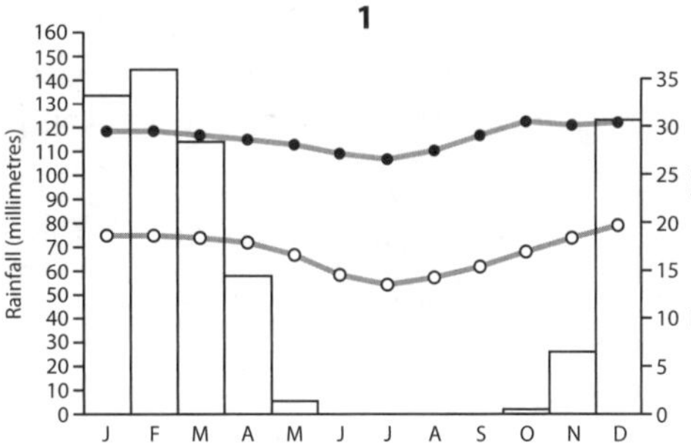
1
Rainfall (millimetres)
Temperature (°C)
J F M A M J J A S O N D

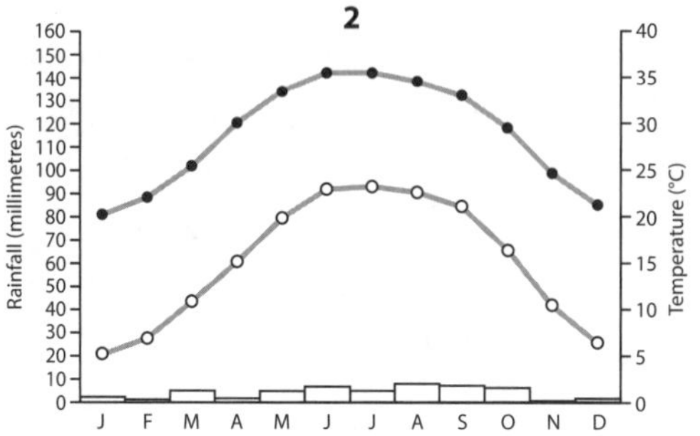
2
Rainfall (millimetres)
Temperature (°C)
J F M A M J J A S O N D

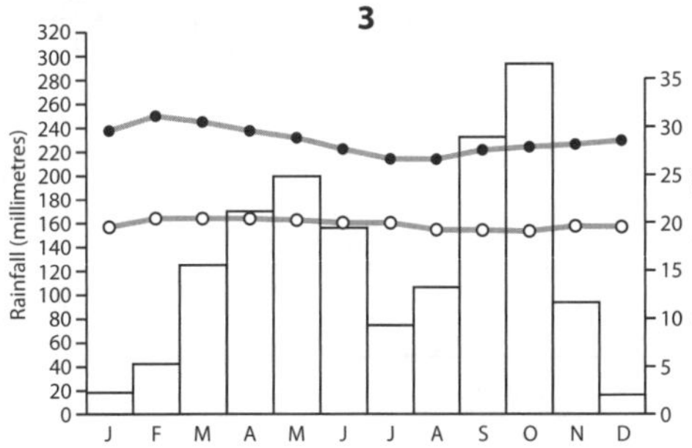
3
Rainfall (millimetres)
Temperature (°C)
J F M A M J J A S O N D

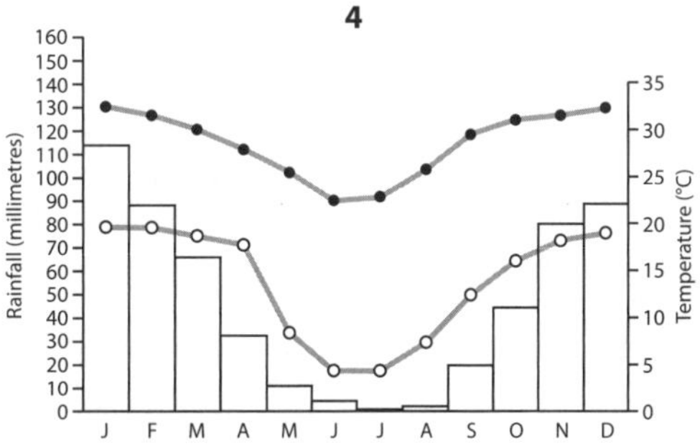
4
Rainfall (millimetres)
Temperature (°C)
J F M A M J J A S O N D

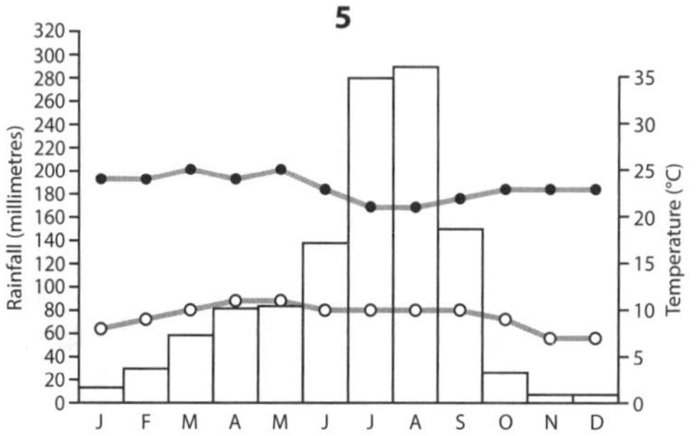
5
Rainfall (millimetres)
Temperature (°C)
J F M A M J J A S O N D

Place A (elevation 1320m) Chart number ________ Name ________________ Biome ________________

Reasons ____

Place B (elevation 2355m) Chart number ________ Name ________________ Biome ________________

Reasons ____

Place C (elevation 726m) Chart number ________ Name ________________ Biome ________________

Reasons ____

Place D (elevation 1120m) Chart number ________ Name ________________ Biome ________________

Reasons ____

Place E (elevation 983m) Chart number ________ Name ________________ Biome ________________

Reasons ____

7 In the Horn of Africa

geog.1 pages 116-117

1 The Horn of Africa is a region of north-east Africa. Use an atlas to find out the closest non-African country to the Horn.

a What is the name of this country? ______________________________

b Which country in the Horn is this country closest to? ______________________________

c Approximately how far apart are the two countries? ______________________________

2 Use the pupil book and your own research to complete the following sentences. Cross out the wrong word from each pair.

The Horn of Africa is a *country / region* in the *north east / north west* of Africa. In total, *five / four* countries make up the region. Much of the coastline is along the *Pacific / Indian* Ocean which forms the coast of *Ethiopia / Somalia*. Many of the people who live in the Horn are *tourists / nomads* whose main work is *farming / mining*. *Djibouti / Addis Ababa* is the largest city in Ethiopia with a population of *4.2 / 6.8* million people.

3 Use the globe below to help you describe the location of the Horn of Africa. Write your answers around the globe.

Tip: Remember to use as many geographical words as you can. **!**

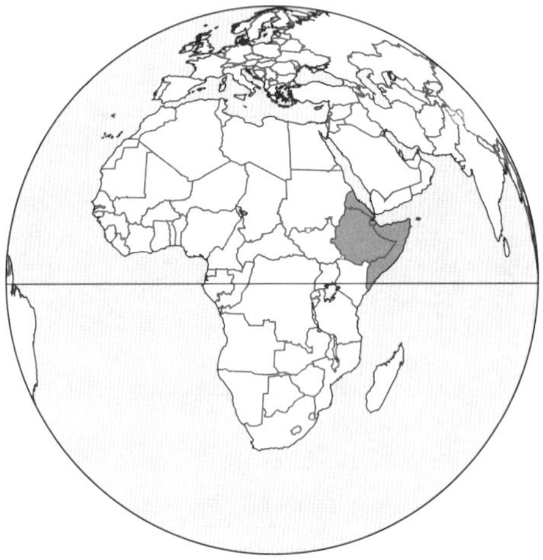

____

____

____

____

____

____

____

4 How would you get the Horn of Africa from Southampton in the UK if you were travelling by sea? Use an atlas to plan a route. Describe where your ship would sail.

Tip: Remember to use geographical words and the names of countries in your answer! **!**

____

____

____

____

7.1 Meet the Horn of Africa

Now you will find out more about the lives of people who live in the Horn of Africa.

4th edition
geog.1

pages 118-119

1 Many of the people who live in the Horn live in straw and mud huts, as shown in the photograph on page 119 in the pupil book. Here is a description of the huts...

The people in this part of Eritrea are pastoral farmers, and keep animals. The people live in round huts made from an interweaving of rods and twigs covered with clay. The thatched, cone-shaped roofs go right down to the ground, making the hut look a little like a 'beehive'. Inside there are mats, often made of woven goat hair. The huts often have two entrances and are usually found in clusters away from main tracks, surrounded by low fences. It is often quite a walk to find water.

In the space below make a drawing of what you think the huts will look like. Use the description to help you. Remember to add some labels.

2 Imagine you live in one of these huts with your family. Write a description of what a typical day might be like. Particularly focus on the adjectives that you use and underline the three best adjectives in your answer. Compare your answer with a partner.

7.2 The Horn of Africa: physical features

This is where you will get physical with this part of Africa!

geog.1

pages 120-121

1 The main physical features of the Horn are listed below. Rank them 1-4, with 1 being your first choice. Which area of the Horn that you would most like to visit? Give two reasons for your answer.

Ethiopian Highlands ☐ Afar Triangle ☐ Ogaden ☐ Coast ☐

Reason 1 ______________________________

____

Reason 2 ______________________________

____

2 Write a fifty word radio advert that would 'sell' your choice to tourists and visitors. Remember your use of adjectives and geographical vocabulary. And it must be *exactly* 50 words!

____

____

____

____

____

____

3 Research your chosen area and create a fact file. Write your answers in the boxes below.

Fact 1:

Fact 6:

Fact 2:

My area is ______________________________

Fact 5:

Fact 3:

Fact 4:

7.3 The Horn of Africa: climate

This is about climate and how it affects farmers and their families.

geog.1
pages 122-123

1 Look at the climate facts below. Next to each, write one effect that each would have on farmers.

Climate fact	Effect
The lower land is hot all year.	
Rain does not fall steadily through the year.	
Some areas have two dry seasons each year.	
In many areas July temperatures exceed 30°C.	
During some years the rains fail completely.	

2 You have been asked to write a summary of the climate in the Horn of Africa for a geography website. Use maps A and B on page 122, and the map on page 118, in the pupil book to help you.

> Tip: Remember to use geographical terms, compass directions and also country names in your answer!

3 Imagine you are a farmer in the Horn of Africa. Describe where you think it would be best to farm and give reasons for your answer. Write your answer in the speech bubble below.

7.4 Coffee farming in Ethiopia

geog.1

pages 124-125

This is about how Ethiopian coffee farmers try to cope with the challenges they face.

Oromia Coffee Farmers' Co-operative Union was founded in 1999. Its members are local farmers in southern and south west Ethiopia which produces two-thirds of the country's coffee. The farms are located in mountainous, rainforest areas at altitudes of 1,500 to 2,000 metres where electricity and clean running water are rare. The co-operative exports and sells coffee on behalf of the farmers and some of the profits are then used to support social projects such as schools and health care. It aims to help its members to become economically self-sufficient and make sure that families can feed themselves if harvests fail, which could lead to famine. It also aims to help farmers cope better with changes in world coffee prices that can affect their income.

Coffee is a globally traded commodity just like oil. Prices go up and down because of factors such as changing weather conditions in the major producing countries, political unrest, worries about now much may be grown (too much or too little), changing transport costs (which is influenced by the number one commodity, oil!) and other unexpected factors. These may include news of a possible drought or freezing conditions in coffee producing areas which would likely mean that there is less coffee available globally. Prices would then go up.

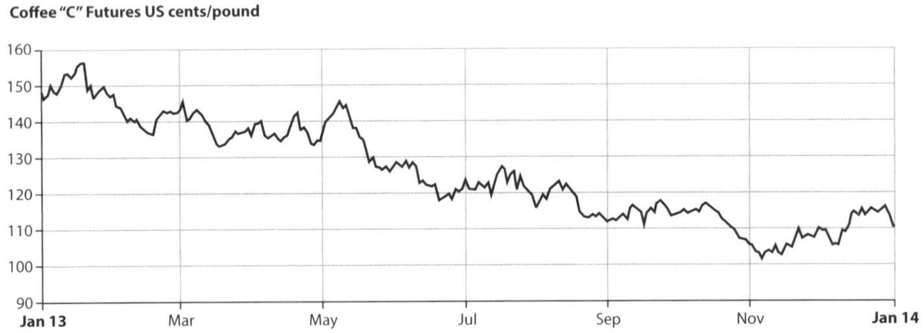
Coffee "C" Futures US cents/pound
160
150
140
130
120
110
100
90
Jan 13
Mar
May
Jul
Sep
Nov
Jan 14

1 Read the information above and study the graph of world coffee prices. Write down one farming challenge, and one economic challenge, that Ethiopian coffee farmers face.

Farming challenge: ______________________________

____

Economic challenge: ______________________________

2 Write a description of what has happened to world coffee prices during 2013. What problems would these bring to both coffee farmers and the Oromia Cooperative?

____

____

____

3 Find out how much of your own family's shopping is Fair Trade. Are there things that you buy that are not, but could be? Make a family shopping list of Fair Trade items.

____

____

4 Look at where the items on your shopping list have come from. Mark the countries on a world map. What do you notice about where your items come from? Compare your answer with a partner; do you notice any patterns?

7.5 Life as a nomad

This is about the nomads, who live and travel with their animals.

geog.1 pages 126-127

1 Complete the following statements, using the words from the box.

- Nomads are people who rear ______________ .
- Nomads live in the areas of the Horn of Africa ______________ .
- Nomads travel with their animals to find ______________ .
- Rivers that flow in the rainy season are called ______________ .
- Land that is not good to grow crops on is called not ______________ .
- The homes of many Nomad's move; the homes are ______________ .

grazing portable fertile animals seasonal dry

2 Look at the photograph below showing nomads at a well. Imagine you are there as an investigative journalist for a UK newspaper. You have been asked to write about the lives of the nomads.

Write a report of your visit: what are the people doing? Write down what you see, hear and smell.

3 What do you think is the greatest challenge that nomads face in their daily lives? Write your answer below and give reasons for your choice.

7.6 Working as a salt miner

geog.1

pages 128-129

This is about the future for the Danakil depression, one of the world's most hostile environments.

1 Dallol, a remote mining camp only accessible by camel, is found in the Danakil Depression. Read these facts and then tick the correct box next to them. Does the fact refer to the past, the present or the future?

Tip: You can tick more than one box!

	Past	Present	Future
It was once a busy site mining potash, sylvite, and salt.	☐	☐	☐
In 1906, when the salt was discovered, an Italian mining company built a railway to the coast.	☐	☐	☐
Getting to Dallol from any direction is long and difficult.	☐	☐	☐
The locals refer to the area as "the Gateway to Hell."	☐	☐	☐
China and India have agreed to help Ethiopia build 5,000 km of railway giving access to the Danakil potash region.	☐	☐	☐
Thin crust covers pools of acid, bubbling to the surface through hot springs and geysers which spit out toxic gases.	☐	☐	☐
Djibouti has close proximity to India, the second largest importer of potash.	☐	☐	☐
There are currently no medical facilities in the region.	☐	☐	☐
The harbour facilities in Djibouti are being upgraded.	☐	☐	☐
At Dallol, salt deposits reach the earth's surface, which made mining more profitable.	☐	☐	☐
Today the town has long been abandoned, only parts of the salt-block walls of buildings remain.	☐	☐	☐

	Past	Present	Future
The journey by camel can take a full day from the nearest village.	☐	☐	☐
Allana Potash plans to spend $642m in three years on four projects in Ethiopia's northeast.	☐	☐	☐
A Canadian company, Allana Potash, was granted a mining permit on 9 October, 2013.	☐	☐	☐
There is the threat of "fire wind," the hot sandstorms which have been described as feeling like a 'tornado in an oven'.	☐	☐	☐
In 2011, the global demand for fertilizer totalled 185 million tonnes, with over 50 million tonnes being potassium-based.	☐	☐	☐
Two communities in the area are being resettled as a health and safety precaution in early 2014.	☐	☐	☐
Potash is a potassium-based mineral used in fertilizer.	☐	☐	☐
A new road will cross the Danakil basin, an important link for importing equipment and the export of the potash.	☐	☐	☐
A tourist accidently fell into a natural sulphur spring at Mount Dallol, a popular tourist attraction, suffering burns to 35% of the body.	☐	☐	☐

2 What do you think the future holds for Dallol – should the company be allowed to mine?

If you were in charge of planning for the region's future, what decisions would you make? Why? What impact will your decision have, and on who?

____

____

____

____

7.7 Life on the coast

This is about the opportunities for tourism in the Horn of Africa.

4th edition
geog.1

pages 130-131

1 The countries in the Horn of Africa have many physical (natural) advantages. Identify three of these advantages and say why they could be attractive to tourists.

Physical (natural) advantage	Attractive to tourists because...

2 Read the extract below.

Mo Farah was born in Mogadishu in Somalia in 1983. His double Olympic gold gives a much-needed positive image for Somalia whose name has unfortunately become linked with anarchy and lawlessness. In 2014 the UK and USA Governments both still advised against travelling to the country.

However, Somalia has many features that could prove attractive to tourists, not least its Indian Ocean white sandy beaches. Lido Beach, Mogadishu's most popular stretch of coast, offers visitors a beach scene that is common in much of Europe and beyond – ice cream, cold drinks, and children playing in the sea whilst young adults play football. There are street cafés and fruit sellers – all seemingly representing the 'typical' tourist destination. Tourism is one area of the economy that Somalia is very keen to develop.

Look at the list below showing some of the things people consider when deciding where to go on holiday. Circle each 'emoticon' to show the extent to which you think Somalia meets these expectations.

Tip: You may need to do some research as well as using the pupil book to help you!

Wide range of activities	☹	😐	☺
Ease of getting there	☹	😐	☺
Interesting history	☹	😐	☺
Good hotels	☹	😐	☺
Reputation of the tourist destination	☹	😐	☺
The weather	☹	😐	☺
The food	☹	😐	☺
Safety	☹	😐	☺

3 What do you think Somalia should do to encourage more tourism? Prepare a five point plan that would help them.

7.8 In the city: Addis Ababa

This is where you find out more about Ethiopia's capital city.

4th edition
geog.1
pages 132-133

1 Cross out the wrong word or phrase in each sentence.

- Addis Ababa is home to the *European Union / African Union*.
- You will see many fine buildings, shopping centres and *old / modern* blocks of flats.
- A lot of people are unemployed, over a *quarter / half* of the workforce.
- Many of the *slums / houses* in the city are rented out by the government at a very low rent.
- The population of Addis Ababa is growing *fast / slowly*.

2 Here are some facts about Addis Ababa. Colour the one that you think proves most challenging for the city to deal with. Write three ways in which the city could cope with the issue.

Addis Ababa

- A birth rate of 23 for every thousand people
- The average life expectancy is around 65 years
- The population will double about every 33 years
- 50% of the people live in poverty
- 20% of the population are under 15 years old
- Its population of 4.2 million grows by 140 000 each year
- 39% of the population can read and write
- Over 25% of the population have no job

Way 1 ______________________________

Way 2 ______________________________

Way 3 ______________________________

3 You are lucky - you have been awarded a new flat in the lottery! Write an email to your family in the countryside explaining how you feel about the news.

7.9 Djibouti: a great location

Here you can learn about some of the difficult dilemmas that Djibouti faces.

4th edition
geog.1

pages 134-135

May 2012: After six years of consecutive drought, Djibouti faces severe food insecurity. Food production from both crops and livestock remains extremely poor. Many rural households have migrated within their region or moved into the towns looking for work. Households unable to afford to move have suffered serious livestock loss and the amount of farmed land has dropped sharply. As more than 90 per cent of food is imported, the country is highly susceptible to global price changes. The poor, who spend up to three quarters of their income on food, are particularly vulnerable to high food prices. (Adapted from http://www.wfp.org/countries/djibouti)

August 2013: the United Nations High Commissioner for Refugees (UNHCR) reported that Djibouti had been receiving refugees and asylum seekers from neighbouring countries for many years. There have been steady arrivals of people fleeing their countries for a combination of reasons, such as war and civil unrest, persecution and poverty.

December 2013: Djibouti's Energy Minister signed an agreement with a Chinese transnational electricity company for the construction of a 90 km power line. The project will help supply energy to the planned 784 km railway line, announced in July 2013, which will link Djibouti to Addis Ababa in Ethiopia.

1 The Djibouti government have to make decisions about where to spend the limited amount of money that the country has. Complete the table below.

Issue	What is the issue?	Is it good or bad for Djibouti? Why?
Issue 1: May 2012 Food insecurity		
Issue 2: Aug 2013 Refugees		
Issue 3: Dec 2013 Electricity line		

2 The government has to prioritise its spending. Where do you think it should spend its money? Who would benefit? Choose one of the issues above. Give reasons for your answer.

I think that issue 1 / 2 / 3 should be the government's priority because ______________________________

____

7.10 How is the Horn of Africa doing?

This is where you will use data to help understand more about the Horn.

4th edition
geog.1

pages 136-137

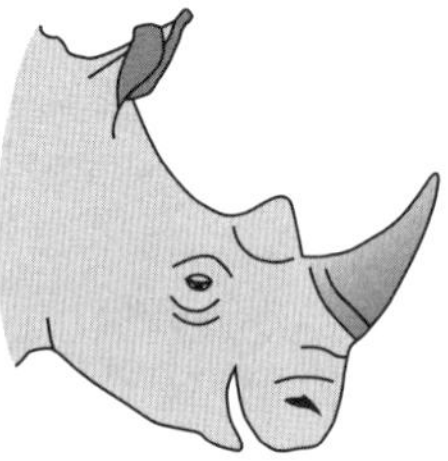

1 Imagine you live in one of the countries in the Horn of Africa. You have been asked to give your five greatest hopes for the region in the future – your own wish list of progress or changes that you would like to see happen. What would you put on your wish list and why? Use data from pages 136-137 to help support your answer.

Wish 1 ______________________________

____

____

Wish 2 ______________________________

____

____

Wish 3 ______________________________

____

____

Wish 4 ______________________________

____

____

Wish 5 ______________________________

____

____

2 Now think about everything that you have learned about the Horn of Africa. Draw a picture that represents what you understand the Horn to be like. Use labels and try to be as creative as possible!